T0334593

Energy for Sustainable Development

Energy for Sustainable Development

Demand, Supply, Conversion and Management

Edited by

Dr. Md. Hasanuzzaman
Higher Institution Centre of Excellence (HICoE),
UM Power Energy Dedicated Advanced Centre (UMPEDAC),
University of Malaya, Jalan Pantai Baharu,
Kuala Lumpur, Malaysia

Dr. Nasrudin Abd Rahim
Higher Institution Centre of Excellence (HICoE),
UM Power Energy Dedicated Advanced Centre (UMPEDAC),
University of Malaya, Jalan Pantai Baharu,
Kuala Lumpur, Malaysia

Academic Press is an imprint of Elsevier
125 London Wall, London EC2Y 5AS, United Kingdom
525 B Street, Suite 1650, San Diego, CA 92101, United States
50 Hampshire Street, 5th Floor, Cambridge, MA 02139, United States
The Boulevard, Langford Lane, Kidlington, Oxford OX5 1GB, United Kingdom

Notices
Knowledge and best practice in this field are constantly changing. As new research and experience broaden our understanding, changes in research methods, professional practices, or medical treatment may become necessary.

Practitioners and researchers must always rely on their own experience and knowledge in evaluating and using any information, methods, compounds, or experiments described herein. In using such information or methods they should be mindful of their own safety and the safety of others, including parties for whom they have a professional responsibility.

To the fullest extent of the law, neither the Publisher nor the authors, contributors, or editors, assume any liability for any injury and/or damage to persons or property as a matter of products liability, negligence or otherwise, or from any use or operation of any methods, products, instructions, or ideas contained in the material herein.

Library of Congress Cataloging-in-Publication Data
A catalog record for this book is available from the Library of Congress

British Library Cataloguing-in-Publication Data
A catalogue record for this book is available from the British Library

ISBN: 978-0-12-814645-3

For information on all Academic Press publications visit our website at https://www.elsevier.com/books-and-journals

Publisher: Joe Hayton
Acquisition Editor: Lisa Reading
Editorial Project Manager: Lindsay Lawrence
Production Project Manager: Selvaraj Raviraj
Cover Designer: Christian J Bilbow

Typeset by TNQ Technologies

Contents

4. Energy supply

5. Energy demand forecasting

6. Energy storage technologies

Fayaz Hussain, M. Zillur Rahman, Ashvini Nair Sivasengaran and M. Hasanuzzaman

7. Energy economics

Laveet Kumar, M.A.A. Mamun and M. Hasanuzzaman

Contributors

Hang Seng Che, Higher Institution Centre of Excellence (HICoE), UM Power Energy Dedicated Advanced Centre (UMPEDAC), Level 4, Wisma R&D, University of Malaya, Jalan Pantai Baharu, Kuala Lumpur, Malaysia

M. Hasanuzzaman, Higher Institution Centre of Excellence (HICoE), UM Power Energy Dedicated Advanced Centre (UMPEDAC), Level 4, Wisma R&D, University of Malaya, Jalan Pantai Baharu, Kuala Lumpur, Malaysia

Fayaz Hussain, Higher Institution Centre of Excellence (HICoE), UM Power Energy Dedicated Advanced Centre (UMPEDAC), Level 4, Wisma R&D, University of Malaya, Jalan Pantai Baharu, Kuala Lumpur, Malaysia

Nur Iqtiyani Ilham, Higher Institution Centre of Excellence (HICoE), UM Power Energy Dedicated Advanced Centre (UMPEDAC), Level 4, Wisma R&D, University of Malaya, Jalan Pantai Baharu, Kuala Lumpur, Malaysia; Faculty of Electrical Engineering, Universiti Teknologi MARA (UiTM), Masai, Johor, Malaysia

M.A. Islam, Higher Institution Centre of Excellence (HICoE), UM Power Energy Dedicated Advanced Centre (UMPEDAC), Level 4, Wisma R&D, University of Malaya, Jalan Pantai Baharu, Kuala Lumpur, Malaysia

M.M. Islam, Higher Institution Centre of Excellence (HICoE), UM Power Energy Dedicated Advanced Centre (UMPEDAC), Level 4, Wisma R&D, University of Malaya, Jalan Pantai Baharu, Kuala Lumpur, Malaysia

Laveet Kumar, Higher Institution Centre of Excellence (HICoE), UM Power Energy Dedicated Advanced Centre (UMPEDAC), Level 4, Wisma R&D, University of Malaya, Jalan Pantai Baharu, Kuala Lumpur, Malaysia

M.A.A. Mamun, Higher Institution Centre of Excellence (HICoE), UM Power Energy Dedicated Advanced Centre (UMPEDAC), Level 4, Wisma R&D, University of Malaya, Jalan Pantai Baharu, Kuala Lumpur, Malaysia

A.K. Pandey, Higher Institution Centre of Excellence (HICoE), UM Power Energy Dedicated Advanced Centre (UMPEDAC), Level 4, Wisma R&D, University of Malaya, Jalan Pantai Baharu, Kuala Lumpur, Malaysia

N.A. Rahim, Higher Institution Centre of Excellence (HICoE), UM Power Energy Dedicated Advanced Centre (UMPEDAC), Level 4, Wisma R&D, University of Malaya, Jalan Pantai Baharu, Kuala Lumpur, Malaysia

M. Zillur Rahman, Department of Industrial Engineering, Faculty of Engineering, BGMEA University of Fashion and Technology (BUFT), Dhaka, Bangladesh

Ashvini Nair Sivasengaran, Higher Institution Centre of Excellence (HICoE), UM Power Energy Dedicated Advanced Centre (UMPEDAC), Level 4, Wisma R&D, University of Malaya, Jalan Pantai Baharu, Kuala Lumpur, Malaysia

Yuan Yanping, School of Mechanical Engineering, Southwest Jiaotong University, Chengdu, China

Biography

Dr. Md. Hasanuzaman received PhD and M. Eng. Sc. from University of Malaya, Malaysia and BSc in Mechanical Engineering from Bangladesh University of Engineering and Technology (BUET), Bangladesh. Dr. Md. Hasanuzaman is currently a Senior Lecturer and Program Coordinator (Master of Renewable Energy) in the UM Power Energy Dedicated Advanced Centre (UMPEDAC), Higher Institution Centre of Excellence (HICoE), University of Malaysia. Presently he is Associate Editor in Chief and Editorial Board member of several international journals. He works in the field of Thermal Engineering, Energy Conversion, Renewable Energy, Energy Policy, Solar Energy, Photovoltaic, PV/Thermal, Solar Heat in Industrial Process, Energy and Buildings, Transport and Electric Vehicle, and Photovoltaic Materials. He received University of Malaya Excellence Awards 2012 for his outstanding achievement in PhD, Technical Scholarship from BUET, Bangladesh Scholarship Council, and the Nippon Foundation, Japan, 2003–2004.

Prof. Ir. Dr. Nasrudin Abd Rahim received BSc (Hons) in Electrical Electronic Engineering and MSc in Electrical Power Engineering from University of Strathclyde, Glasgow, UK, in 1985 and 1988, respectively. He received PhD in Power Electronics from Heriot-Watt University, Edinburgh, UK, in 1995. He is currently a Professor at University of Malaya, Kuala Lumpur, Malaysia, and Director/Founder of UM Power Energy Dedicated Advanced Centre (UMPEDAC), a recognized Higher Institution Centre of Excellence (HICoE) under Ministry of Higher Education. He is a Distinguished Adjunct Professor of King Abdulaziz University, Jeddah (2012–2017). His research interests include Power electronics and drives, Solar photovoltaic technologies, Real-time control systems, Transportation Demand Side Management, and Energy Policy, with over 400 technical papers in journal and international conference proceedings. Prof. Nasrudin is a Chartered Engineer (UK), Fellow of the Institution of Engineering and Technology, UK, and Senior Member of the Institute of Electrical and Electronics Engineers, USA. He is also a Fellow of the Academy Science and Professional Engineer, Board of Engineer, Malaysia.

Chapter 1

Introduction to energy and sustainable development

M.M. Islam, M. Hasanuzzaman
Higher Institution Centre of Excellence (HICoE), UM Power Energy Dedicated Advanced Centre (UMPEDAC), Level 4, Wisma R&D, University of Malaya, Jalan Pantai Baharu, Kuala Lumpur, Malaysia

1.1 Energy and civilization

The modern worldview is constructed around the central concept of energy. Energy is a concordant notion that encompasses not only physical and environmental science but also the socioeconomic disciplines. Physicists define energy as the capacity to do work as measured by the capability to do work (potential energy) or the conversion of this capability to motion (kinetic energy). On the other hand, in modern economic concept, energy is considered as fuel, a substance used as source of energy (Martinás, 2005). Energy is of worth if it is easily convertible to useful work, and most of the world's convertible energy comes from fossil fuels. That is why fuel is the most dominant factor apart from food affecting the human civilization.

Unlike all other animal species, human beings derive their energy from both food and fuel. At a very early stage of civilization, human beings used tools and weapons to effectively apply their somatic strength, but these were not the sources of energy. The maiden access to a source of energy was possible when they learnt to control the fire. Later, man tried to control the energy of water and wind. However, the epochal discovery of fossil fuels brought the major impact in every sector of human civilization, from agriculture to industry, from economics to politics, and most importantly the environment. The great ages of civilization are characterized in terms of new source or form of energy used or introduced in that era, e.g., fossil fuel era, post fossil fuel era, nuclear era, era of renewables, etc. Historical industrial revolutions were essentially energy revolutions. The first industrial revolution (1760–1840) saw the advent of mechanization that led to the formation a new economic structure. Steam engines powered by coal brought about revolutions in production system and communication network; handlooms were replaced by power looms, railway and steamboats brought the far near. With the

Energy for Sustainable Development. https://doi.org/10.1016/B978-0-12-814645-3.00001-8

emergence of a new source of energy, i.e., electricity, gas, and oil, the second industrial revolution instigated at the late 19th century (1870) and extended up to the World War I. This was essentially a technological revolution and connoted by the electrification of manufacturing systems that made possible the assembly line and mass production (Ford & Crowther, 1922). The third industrial revolution started (1969) as nuclear power generation commenced. Prodigious development in electronics, telecommunication, and computing technology, especially programmable logic controllers and robots, led to the automation of the production system. The first industrial revolution mechanized the production system through the use of steam power, electric power facilitated mass production in the second industrial revolution, and industrial automation came into reality in the third industrial revolution because of electronics and information technology (Schwab, 2016). A fourth industrial revolution is building on the third and integrating the production processes with alternative energy resources such as wind, sun, and geothermal energy (Sentryo, 2017).

Today, energy is directly related to the most critical economic and social issues that affect sustainable development such as water supply, sanitation, mobility, food production, environmental quality, education, job creation security, and even peace in global context. Energy nowadays has become the universal currency. While extra energy at the disposal of the first world countries is widening their power and control, access to the energy sources is becoming limited to the under developed countries. The global economic disparity among the nations is predominantly perceived by their respective control over energy and fuel sources. It has been estimated that about 1.2 billion people are out of electrical network and more than 2.0 billion people worldwide are deprived of mere access to modern energy resources (Behl, Chhibar, Jain, Bahl, & El Bassam, 2013). The competition has gone so hostile that it led to series of "oil war" including two gulf wars in the Middle East region. This raised the question how to ensure a more equitable energy dissemination and whether it is possible to attain control in the demand for energy. As the former clause is related to complex global politicoeconomic order, researchers have emphasized on the later one, that is, abating the energy demand and at the same time ensuring efficient use of available energy emphasizing on more and more infliction of renewable energy technologies.

1.2 Global energy resources

There are numerous sources of energy all around the nature. The conventional sources of energy can be classified as nonrenewable and renewable. Nonrenewable energy resources, such as coal, nuclear, oil, and natural gas, are limited in reserve and supply because of the very long time required for them to be replenished in nature. On the other hand, renewable resources are replenished continuously or over very short time span. The major renewable energy resources are solar, wind, hydro, biomass, geothermal, and marine or ocean energy (Andrea, 2014). There are some unconventional sources such as waste-to-energy (WtE), carbon capture, and storage, etc., which have been emerging as energy sources in recent years. Nowadays, energy storage and energy efficiency are also considered as energy source.

1.2.1 Coal

Coal is the transformed remains of prehistoric vegetation amassed in swamps and peat bogs. It is an ignitable organic sedimentary rock composed mainly of carbon, hydrogen, and oxygen (WCA, 2009). Coals are ranked based on their physical and chemical properties that have been developed in course of coalification (the process of maturation from peat to anthracite). Soft lignite and subbituminous coals with dull fictile appearance are ranked low-grade coal. The moisture level is high and carbon (thereby, energy) content is low. Coals with higher carbon and lower moisture content are ranked higher (e.g., anthracite) and generally harder and stronger rocklike material, often with black vitreous luster. Fig. 1.1 depicts classification of coals according to their energy and moisture content.

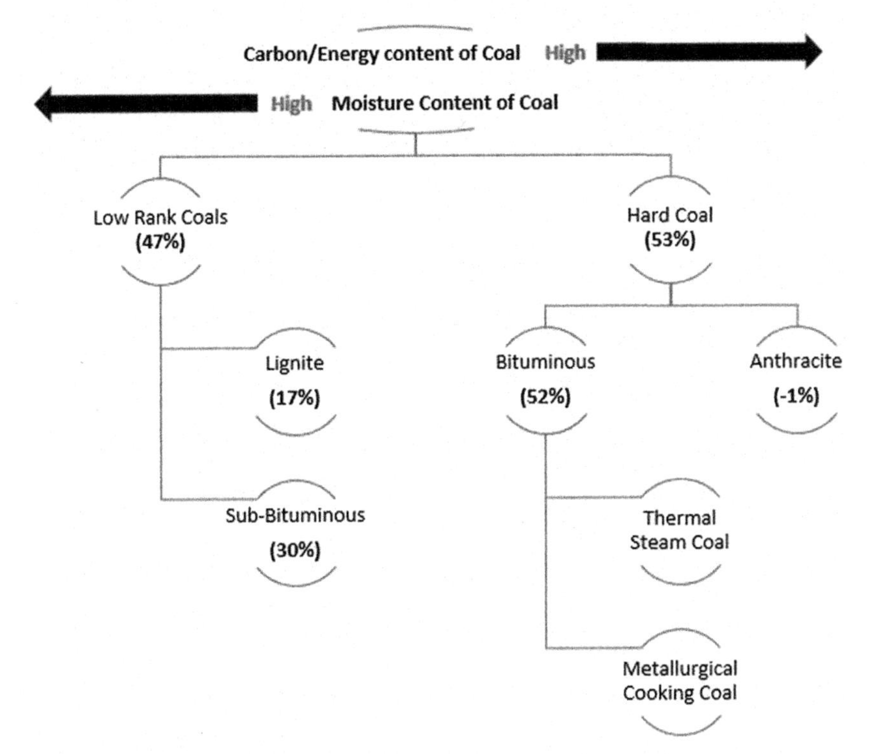

FIGURE 1.1 Classification of coal according to energy and moisture content (Kavalov & Peteves, 2007).

A major share (around 30%) of world's primary energy consumption is met by coal in range of sectors including power generation, iron and steel production, cement manufacturing, etc. Currently, about 40% of global power plants run on coal burning; on the other hand, coal is used in 70% of the steel production and 50% of aluminum production plants (WCA, 2009). From 2000 to 2014, global coal consumption rose by 64%, which portarys its dominating role in world economy (WEC, 2016). Distribution of world proven coal reserves in 2016 is shown in Fig. 1.2.

Coal is the most carbon-intensive fossil fuel, and its undeterred use in power generation could have significant impact on global climate change. However, modern carbon capture utilization and storage technology allows higher efficiency and low carbon emissions of coal power plants, still upholding its utility in energy sector.

1.2.2 Oil

Oil is liquid mixture of hydrocarbons found in suitable rock strata and can be extracted and refined to produce fuels including petrol, paraffin, and diesel oil. Different grades of petroleum oils (molecular weight of the hydrocarbon ranging from 1 to 70) are produced from crude by factional distillation method. In oil industry, oil (also raw natural gas) extracted by natural pressure after drilling is known as conventional oil (and conventional gas); on the other hand, unconventional oil is extracted by means of techniques other than traditional method. Petroleum product derived by unconventional methods

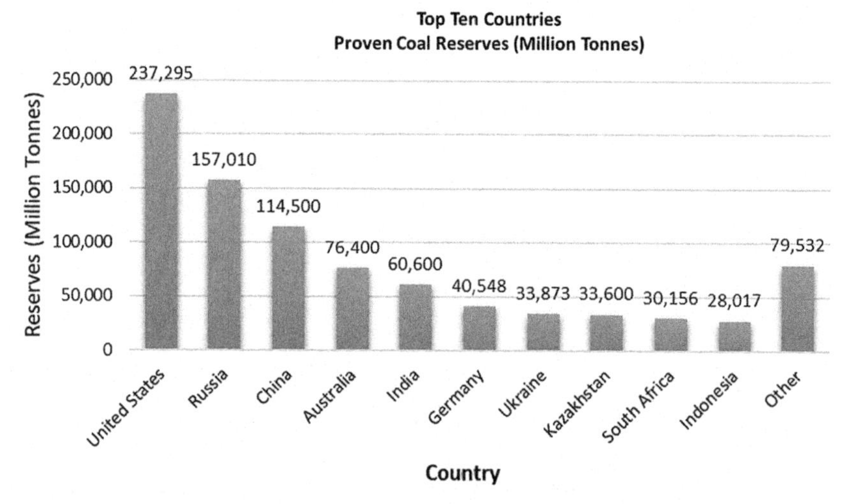

FIGURE 1.2 Distribution of proven coal reserve in 2016. *BP Statistical review of world energy 2016.*

from fine-grained, organic-rich rocks (known as oil shale) is called shale oil (and gas) or light tight oil. The estimated global shale oil resource is around 6050 billion barrels (bbl), which is four times the world's conventional crude oil reserve (WEC, 2016).

There is no major chemical difference between unconventional and conventional oil and gas. Crude oil quality depends on its sulfur content, which is an extremely corrosive element and produces toxic hydrogen sulfide gas. Crude oil with total sulfur level over 0.5% is called "sour" and oil with lower sulfur is termed "sweet." Light (lower density) and sweet (low sulfur content) crude oil are usually costlier than heavy, sour oils because light sweet grades can be processed with less sophisticated and energy-intensive refineries.

Oil accounted for 32.9% of total global energy consumption in 2016, and from 2014 to 2016, world oil production was increased by 3.0%. In the last 20 years, proven crude oil reserves have grown from 1126.2 to 1697.6 bbl. Fig. 1.3 shows the world proven crude oil reserve by region.

1.2.3 Natural gas

Natural gas is an inflammable mixture of gaseous hydrocarbons containing primarily methane (usually more than 80%) along with small amounts of ethane, propane, butane, and sometimes nitrogen and helium, too. Natural gas reserves are found in porous sedimentary rocks near other solid and liquid hydrocarbon beds. Many by-products, such as propane, ethane, butane, carbon dioxide, nitrogen etc., are extracted along with natural gas mines.

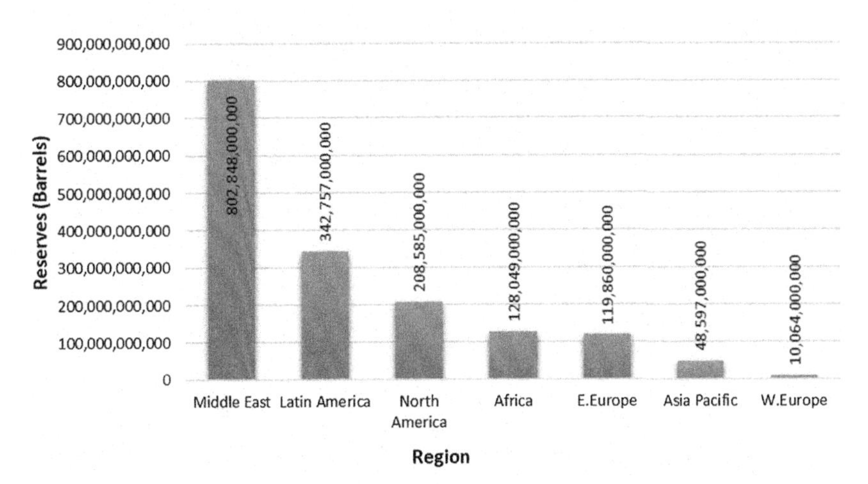

FIGURE 1.3 World proven crude oil reserve by region 2016. *OPEC Annual bulletin 2016, and BP stat review 2016.*

Similar to shale oil, shale gas is available in shale or tight sandstones. Currently only four countries, Canada, the United States, China, and Argentina, are producing commercial volumes of shale gas (EIA, 2015). Another form of natural gas, known as coal mine methane (CMM) or coal bed methane or coal seam gas, came into market as a result of the development of coal mine degasification and safety techniques. CMM is captured from the air in the working coal mines basically to prevent uninhibited release of methane into atmosphere. The proportion of methane in CMM ranges from 25%–60% with an oxygen content of 5%–12% (CE, 2017).

Apart from using it in raw form, natural gas is also stored and used in several other forms such as CNG (compressed natural gas) and LNG (liquefied natural gas). Natural gas compressed to less than 1% of its volume at STP (standard temperature and pressure) is known as CNG. It is generally stored at 20–25 MPa in cylindrical or spherical cylinders and used mainly as a substitute for gasoline in light vehicles. To ensure safe storage and transport, natural gas is converted into liquid (LNG) of pressures near to the atmosphere by cooling it to approximately −162°C, where the volume is condensed to 1/600th the volume of gaseous natural gas at STP. The maximum transport pressure for LNG is kept around 25 kPa (Shaw, 2013).

Natural gas is the lowest carbon-emitting fossil fuel that can serve as a source for base load power at affordable price. With a share of 22%, it is in the second position to fuel power sector and in the third place in terms of primary energy source fulfilling the 24% of global primary energy supply (WEC, 2016). Distribution of worldwide proven gas reserve is shown in Fig. 1.4.

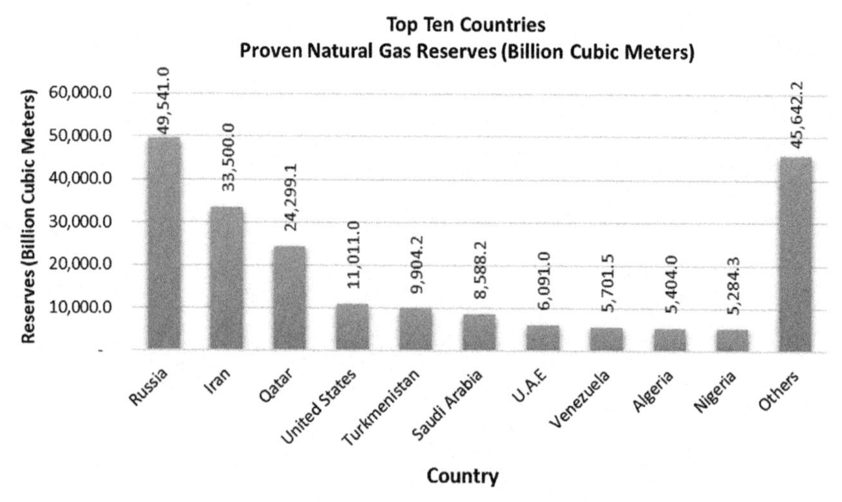

FIGURE 1.4 Distribution of worldwide proven gas reserve. *2016 OPEC Annual Statistical Bulletin.*

1.2.4 Nuclear energy

Nuclear energy is the energy that holds neutrons and protons together in the nucleus of an atom and can be released in the form of heat through nuclear fusion or nuclear fission reactions. This enormous heat energy is used to generate steam in the boiler of a steam turbine and thus produce electricity. Nuclear fission can be natural (known as radioactive decay) or be initiated inside nuclear reactors by splitting the nucleus of the atoms of fissile materials. The most commonly used fissile material in commercial nuclear power generation is the isotope U-235 of uranium.

There reactor technologies developed to run and manage fission reactions are pressurized water reactor (PWR), boiler water reactor (BWR), pressurized heavy water reactor, gas-cooled reactor, liquid water cooled graphite moderated reactor, fast breeder reactor, high-temperature gas-cooled reactor, etc. More than 90% of the operational nuclear power plants are light water reactor designs (PWR or BWR) that run between 50 and 60 years. It is estimated that about 11% of the world's electricity generation is from nuclear plants, and in 2016, nuclear plants supplied 2476 TWh of global power (WNA, 2017).

Nuclear power is gradually being accepted worldwide as a reliable and secure source of power. Although these plants yield radioactive wastes, the volumes are small and can be disposed safely. Rather, with zero carbon emissions, its environmental footprint is trivial.

1.2.5 Hydropower

Hydropower is power derived from the force of falling or fast moving water utilizing the natural water cycle. Water power has long been used in pumping and irrigation and operating devices such as sawmills, textile mills, trip hammers, dock cranes, etc. There are two basic types of water turbines: reaction turbine and impulse turbine. Francis turbine invented by James Francis in 1849 is a mixed flow reaction turbine. It is the first modern water turbine to be employed for power generation and still is the most widely used turbine in hydroelectric power plants around the globe. Kaplan turbine is also a reaction type turbine with an axial-flow design and suitable for the sites where the available head is low. Pelton wheel is the only impulse type water turbine, and this design is basically suited for very high head applications. Hydropower plants are of four types: run-of-river hydropower, storage hydropower, pumped-storage hydropower, and offshore hydropower. The first commercial hydroelectric power plant was built at Niagara Falls in 1879 (IHA, 2016).

Hydropower alone generates 71% of all renewable electricity in the world. The global installed capacity of hydropower has reached 1246 GW in 2017, with an estimated generation of 4102 GWh (IHA, 2016). Three Gorges Dam, China, is world's largest hydropower with a massive capacity of 22.5 GW, generating an average of 88.2 TWh electricity per year (WEC, 2016).

1.2.6 Bioenergy

Bioenergy refers to the use of organic substances or biomass (that is not being fossilized) solely for energy applications. It includes traditional biomass such as forest and agricultural by-products, fish waste, and waste water sludge as well as modern biofuels. Bioenergy technologies are mature enough to transform biomass into heat and electricity. The two sources of bioenergy supply are forest and agriculture, with woody biomass being a major primary source of energy all around the world. Although at present biomass supplies about 59.2 EJ of energy, which is 10.3% of the global final energy use, it has a potential of providing as much as 150 EJ energy by 2035 (IRENA, 2014). Sweden is a world leader in bioenergy utilization with 34% of its final energy use coming from this renewable source.

Modern biomass technologies include direct heat or derived heat from combined heat and power (CHP) plants and liquid biofuels used to power automobiles and operate boilers to generate electricity. Biomass contributed 96.8% of the derived heat and 96.7% of direct heat generation from all renewable sources in 2014. Bioelectricity is third in position as renewable source of power generation. Bioelectricity generation grew at an annual rate of 8.2% during 2000−14. Total electricity generation, in 2014, from electricity-only and cogeneration (CHP) plants is 493 TWh, where electricity-only plants contributed the major share (about 72%) at a global average efficiency of 32%. CHP plants, on the other hand, provided 154 TWh electricity and 0.72 EJ of bioheat in 2014. The use of liquid biofuels is growing substantially in transportation sector including heavy road, aviation, and maritime vehicles. Global biofuel production reached 126 billion liters in 2014 out of which more than 75% was produced in the United States and Brazil. Other bioenergy products include biogas, pellets, and charcoal. Biogas is essentially a mixture of methane and carbon dioxide produced by anaerobic digestion of organic matter. Global production of biogas in 2014 was 58.7 Nm^3. Pellets, a solid biofuel produced from wood dust and agricultural by-product such as straw, is used for residential heating, steam and electricity generation in service industry, etc. Europe leads the world with around 60% of global pellet production. Partial incineration of biomass gives forth to carbonaceous material called charcoal. Global charcoal production in 2015 was 52.2 million tonnes.

1.2.7 Solar

Solar (Latin of sun) is the primary source of all kinds of energy on the earth. Annual global potential of solar energy is 1575–49,837 EJ, which is several times larger than world energy consumption per annum (UNDP, 2000). A variety of technologies have been developed to harness solar energy in useable forms, such as solar photovoltaic modules (PV) for electricity, solar thermal collector (STC) for heat, and photovoltaic thermal for both heat and electricity. PV modules directly convert sunlight into electricity by means of photoelectric effect using semiconductor materials, mostly silicon. Global installed capacity for PV electricity has reached around 227 GWe at the end of 2015, which is 1% of total electricity consumption. STCs provide heat for household to commercial scale with efficiencies as high as 60%–80%. The heat produced by STC can effectively be used to produce air conditioning using absorption systems, to produce electricity by an organic Rankine cycle, to run a Stirling engine to operate thermoelectric devices, etc. Global solar thermal capacity is growing substantially over the past two decades and reached almost 440 GW_{th} in 2015 when the solar thermal yield was 350 TWh (WEC, 2016).

Concentrated solar power (CSP) systems employ mirrors and/or lens to produce intense heat, which is used to generate steam and operate steam turbines for producing electricity. Worldwide CSP capacity has reached to 4815 MW in 2016, which was only 354 MW in 2005 (WEC, 2016). Solar energy is an indigenous, inexhaustible, and mostly import-independent energy resource available in almost all parts of the world. Increased use of solar technologies will help to enhance sustainability and energy security and most importantly reduce pollution.

1.2.8 Geothermal

Geothermal energy is derived from the heat that is produced and stored in subterranean earth through continuous decay of radioactive particles. Geothermal energy is classified as renewable energy (though it has small carbon footprint) because it is continuously recuperated by the heat from the earth's core. About 43×10^6 EJ thermal energy is stored down to 3 km within continental crust, which is far more than the global primary energy consumption (approximately 600 EJ). Geographically, 72% of the geothermal power plants are installed along tectonic plate boundaries of the Pacific Rim and 20% within the Atlantic Basin (Harvey, Beardsmore, Moeck, & Rüter, 2016).

World geothermal electricity generation reached 75 TWh in 2015, and the installed capacity in 2016 reached 13.4 GW. The global geothermal power generation is expected to reach about 18.4 GW by 2021 and 32 GW by 2030s (Matek, 2016). Although overall geothermal share in world electricity generation is still very trivial (only 0.3%), in certain countries such as Kenya (44%

of total generation), Iceland (27%), El Salvador (26%), and New Zealand (18%), it plays the major role in their power sector (WEC, 2016).

1.2.9 Wind energy

The kinetic energy of wind caused by uneven heating of the atmosphere by the sun, irregularities of the earth's surface, and rotation of the earth is a promising source of renewable energy. Electricity produced by wind turbines is called wind power and categorized as large wind onshore, small wind onshore, and offshore turbines. Wind turbines are progressively getting bigger and more powerful. The currently largest available wind turbine is rated at 8.0 MW with a 164 m rotor. The power rating is expected to reach 10–15 MW by 2030s. In end of 2017, wind power capacity reaches 539 GW, providing nearly 4% of total global power generation. Denmark is the leader in wind power with 42% of its electricity generated from wind (WWEA, 2018).

1.2.10 Marine energy

Marine energy (or ocean energy) encompasses wave, tidal stream, tidal range, ocean thermal, ocean current, run-of-river, and salinity, etc., through which energy can be harnessed from oceans. Oceans are the source of enormous untapped energy that is accessible to most coastal countries. The estimated marine energy potential is 32 PWh/y; however, only a minor quantity is harvested worldwide. At present, global marine energy generation capacity is only 500 MW, most of which comes from tidal range (495 MW) along with a small portion from tidal stream (11 MW) and wave (2 MW) (WEC, 2016). Fig. 1.5 shows the global marine energy capacity from 2009 to 2016 (MW) (Sattista, 2017).

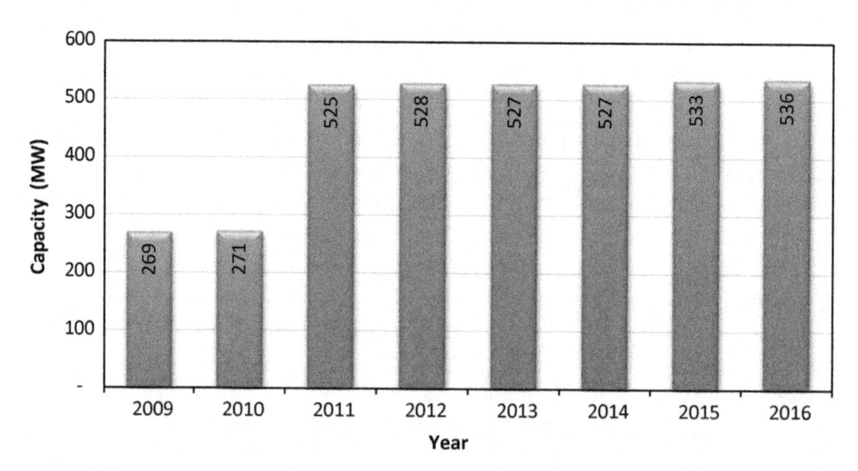

FIGURE 1.5 Global marine energy capacity from 2009 to 2016 (Sattista, 2017).

Marine energy is a geographical location–dependent resource. Moreover, technologies are still not matured, and installation and maintenance costs are quiet high. In addition, such installations cause disturbance in marine ecological system and disruption of natural movement of water.

1.2.11 Waste-to-energy

One of the biggest challenges in forthcoming times is to manage the mammoth piles of waste in a sustainable manner. By 2025, global waste generation crosses more than 6 million tonnes per day. At present, China alone produces about 300 million tonnes of municipal solid waste (MSW) per year, and by 2025, it will be over 500 million tonnes (WEC, 2016).

Considerable amount of energy can be recovered from MSW, construction waste, biowaste, medical waste, etc., depending on their specific composition and energy content. WtE is the process generating electricity and/or heat from the primary treatment of waste. Incineration of waste is used to produce steam in a boiler that drives steam turbine to produce power. Japan uses up to 60% of its solid waste for incineration. The largest circulating fluidized bed WtE plant in China, which was built in 2012, can processes 800 tonnes of waste per day. The 50 MW WtE incineration plant in Addis Ababa, Ethiopia, will process 350,000 tonnes of waste a year (WEC, 2016).

WtE plants help to shrink the urban landfill volume by 90% and thwart tonnes of carbon dioxide emission thereby. Therefore, WtE technology can solve the challenges of both waste management and energy sectors in a sustainable way.

1.2.12 Carbon capture and storage

Carbon dioxide emission, which is the main component of greenhouse gas (GHG), has become one of the major politicoeconomic issues in last two decades. Along with coal-based steam power plants, manufacturing of cement, steel, pulp, and paper, chemicals and processing of natural gas create substantial amounts of CO_2 emission. About 25% of global CO_2 emissions are due to the cement and other manufacturing plants. Carbon capture and storage (CCS) is an integrated suite of technologies that are devised to capture and systematically dispose harmful carbon to reduce GHG emission. CCS includes the followings three actions: separation and compression of CO_2 from other gases, transport to a suitable storage site, and injecting it into selected rock formations typically 1 km underground (CCSA, 2012).

At present, 22 large-scale CCS plants are in operation around the globe, capable of capturing up to 40 million tonnes of CO_2 per year. It has been estimated that a 6000 million tonnes of CO_2 capture per annum by 2050 will facilitate to dwindle the global temperature by 2°C. Canada is the trendsetter in employing large-scale CO_2 capture technology in the power sector. Boundary Dam Power Station at Saskatchewan built the very first CCS facility in October 2014. On the other hand, Abu Dhabi launched the world's first large-scale CCS project in the iron and steel sector in 2016 (WEC, 2016).

CCS is getting popularity day by day for tackling climate change in an affordable way. Industry already gained the skills and experience to safely deliver CCS that will help to build future carbon-free economy.

1.2.13 E-storage

Energy storage (E-storage) refers to conversion of energy to such forms that can be stored conveniently and economically for future use. Hydroelectric dams (both conventional and pumped storage) are familiar examples of bulk energy storage. Basically, there are five categories of E-storage, viz., chemical, electrochemical, electrical, mechanical, and thermal energy storage. Chemical energy storage system employs hydrogen and synthetic natural gas as secondary energy carriers for large-scale electrical energy storage. In electrochemical storage, chemical energy is converted into electrical energy; at least two reagents undergo a chemical reaction in course of the operation, wherein released energy is delivered as electric current at a defined voltage and time. Examples of electrochemical storage include batteries, electrochemical capacitors, fuel cells, etc. Large batteries are already developed with installed storage capacity of 750 MW. Although, in recent years, lithium-ion technology has got more popularity, sodium–sulfur battery accounts for nearly 60% of stationary battery installations. In the Smart Grid Solar Project of Arzberg, Germany, a complete hydrogen system has been installed with proton exchange membrane electrolyzer (75 kW) that transforms the electricity into hydrogen; a 5-kW fuel cell reconverts hydrogen into electricity again (WEC, 2016).

Electrical energy storage system includes capacitors, supercapacitors, and superconducting magnetic energy storage; on the other hand, mechanical energy storage system comprises of flywheel, pumped hydro storage system, and compressed air energy system. About 95% of global E-storage is in the form of pumped hydro storage. Thermal energy systems stores heat (or cold) in storage medium, the mechanism of heat storage being either sensible or latent. Sensible heat storage media may be liquid (water, molten salt, thermal oil, etc.), solid (stone, concrete, metal, etc.), or liquid with solid filler material (molten salt/stone). The 50 MW Bokpoort CSP plant in North Cape Town, South Africa, includes energy storage in molten salt, equivalent to over 9 h of electricity generation. Latent heat storage employs phase change materials that store huge amount of heat at almost constant temperatures (Guney & Tepe, 2017).

1.2.14 Energy efficiency

Energy efficiency is the portion of total energy input to machine or system that is consumed in useful work and not wasted as useless heat or otherwise. It measures how much energy is used by any system or equipment to provide the desired level performance. Nowadays, energy efficiency is worldwide considered as an energy resource because it is capable of yielding energy and demand savings that can displace electricity generation from primary energy resources. Researchers have pointed out energy efficiency as "hidden fuel" or "invisible power" Energy efficiency is the cheapest energy source as there is no need to produce energy in the first place. The single energy resource possessed by all countries in the world is energy efficiency. It has been reported by International Energy Agency (IEA) that in 2016, the world would have used 12% more energy from energy efficiency as a single energy source because of the improvements achieved in energy efficiency since 2000 (IEA, 2017). In addition to the environmental benefits, energy efficiency helps to strengthen energy security. In recent years, researchers are treating energy efficiency as the "first fuel" and a source in its own right. Realizing the impact of energy efficiency in reducing energy intensity and improving energy security policy makers worldwide planning to integrate it into wider national infrastructure planning considering energy efficiency itself as a key infrastructure. Energy efficiency satisfies the definition of infrastructure used by International Monetary Fund and other economic institutions. Because, similar to other traditionally recognized infrastructures, energy efficiency is long-lasting capital stock, provides inputs to a wide range of goods and services, and disburdens other sectors of the economy. Energy saved is like energy produced, so energy efficiency is globally considered as the most cost effective means to supply energy in a sustainable manner along with reducing GHG emission.

1.3 Energy management, energy policy, and energy strategy

1.3.1 Energy management

Energy management is the usage and application of technology including planning and operation of both production and consumption of energy with a view to enhance energy efficiency of an organization (VDI, 2007, p. 3). The principal objectives of energy management are conservation of resources, saving budget and preventing climate change as well as ensuring easy and ingrained access for all to the energy spectrum. Energy Star, a US Environmental Protection Agency (EPA) program, formulated a guideline for scientific

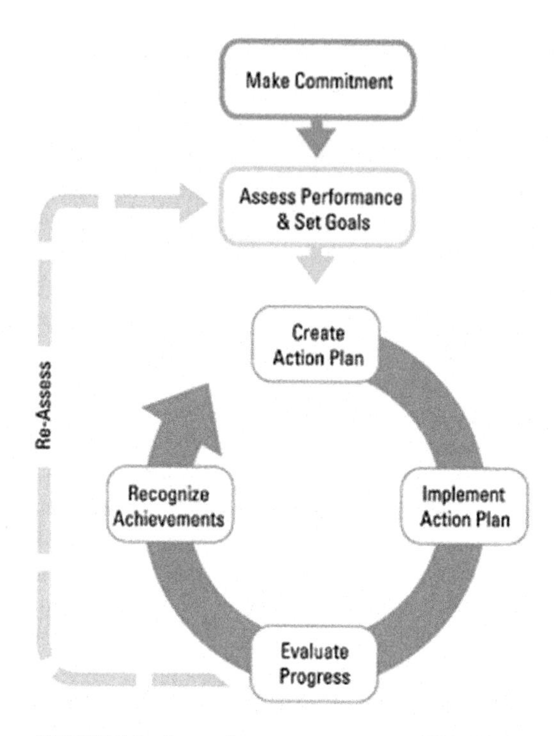

FIGURE 1.6 Steps of energy management (EPA, 2012).

energy management practice; the basic seven steps are as follows (EPA, 2012), which is also depicted in Fig. 1.6.

Step 1: Make commitment
Step 2: Assess performance
Step 3: Set goals
Step 4: Create action plan
Step 5: Implement action plan
Step 6: Evaluate progress
Step 7: Recognize achievements

1.3.2 Energy policy

Energy policy is the scheme in which the government (or any organization) addresses issues related to energy growth and usage including energy production, distribution, and consumption. The attributes of energy policy may include legislation, international treaties, incentives to investment, guidelines for energy conservation, taxation, and other public policy techniques (Armstrong, Wolfram, Gross, Lewis, & Ramana, 2016).

1.3.3 Energy strategy

An energy strategy is a working paper that states how energy will be managed within an organization. A comprehensive energy strategy should address the followings (CT, 2018):

- To ensure the sufficient availability of energy resources and consistent supply.
- To ensure the compliance with energy and climate change regulations.
- To ensure investment in cost-effective energy efficiency projects.
- To ensure the use of energy-efficient equipment and services and control energy costs.

1.4 Energy management for sustainable development

Energy is one of the most central concerns in achieving sustainable development goal. Sustainable development is a lively and mutable process. Along with several other factors such as circuiting investment, adopting appropriate technology, etc., efficient and optimal use of energy is a prerequisite to accomplish sustainable growth. Energy management includes energy saving through introducing energy efficient practice as well as fixing the type of energy and energy tariff appropriate for the respective organization such as power plant, manufacturing factory, commercial and residential buildings, etc. The primary aim of energy management is to ensure optimum energy procurement and utilization as well as to cut down energy cost and reduce energy waste without affecting production and quality with minimal effect on environmental. The engrained relation between energy and the environment has made energy management an unassailable part of modern industrial management system.

One of the basic key activities in energy management is energy audit. It is a systematic approach for those attempts to balance the total energy inputs against its use and serves to identify all the energy streams in a facility along with justifying the energy usage according to its discrete functions. The ultimate goal is to improve energy efficiency, and over a long term this increased efficiency would otherwise lessen energy cost, which in turn will ensure sustainable economic development. In addition, improvement in energy efficiency will ensure energy security along with environmental protection. World economy is expected to almost double over the next 20 years at a rate of 3.4% per annum, whereas the energy demand will increase by 30%. Although global energy consumption rate is expected to drop from 2.2% p.a. in 1995–2015 to 1.3% p.a. in 2016 onwards, civilization still needs huge supply of energy, especially for the developing countries (BP, 2017). Although global energy intensity (amount of energy used per unit of gross domestic product) in 2015 (1.8%) surpassed that attained in 2014 (1.5%), the pace is still too slow to lead the way to a carbon-free economic system. IEA estimates that to achieve global climate goals, annual energy intensity should immediately be raised to at least 2.6% (IEA, 2016). Industrial energy use can be reduced by about 20% through the accomplishment of organizational energy efficiency (IEA, 2013). Moreover, energy efficiency is gradually being recognized as one of the most cost-effective solutions for reducing GHG emissions. It has been assessed that if the energy sector had not been using energy efficient processes since 1973, global energy consumption would have been 50% more than the current level. In addition, if measures on improving energy efficiency are implemented, more than 83 EJ (EJ ~ exajoules $= 10^{18}$ J) energy can be saved by 2030 (Yang & Yu, 2015). Another analysis shows that the energy savings out of increased energy efficiency level (globally 14%) attained from 2000 to 2015 is equivalent to 450 million tonnes of oil equivalent (Mtoe), which saved an expenditure of USD 540 billion. An extra USD 230 billion investment for installing new power plants could be avoided through energy efficiency gains. Most importantly efficiency improvements helped to cut down 1.2 billion tonnes of carbon dioxide (CO_2) emission in 2014, which equals to the total CO_2 emissions in Japan. In 2015, China's achievement in energy savings from efficiency improvement was almost equal to its total renewable energy supply (IEA, 2016).

One of the most important economic issues is the trade-off between energy use and its prices. Monitoring industrial energy performance through the implementation of energy management system can facilitate to enhance productivity and capacity utilization, reduce waste of resources and pollution, and lower operation and maintenance costs. As a result, price of energy and its product will be decreased and value generation will be increased, leading to steady economic growth. The unperturbed economic growth together with environmental protection will form the basis of secure and sustainable development of human society.

References

Andrea, A. (2014). *Nonrenewable and renewable energy resources.* Retrieved from https://ww2.kqed.org/quest/2014/02/13/nonrenewable-and-renewable-energy-resources-2/.

Armstrong, R. C., Wolfram, C., Gross, R., Lewis, N. S., & Ramana, M. V. (2016). The frontiers of energy. *Nature Energy, 1*(11).

Behl, R. K., Chhibar, R. N., Jain, S., Bahl, V. P., & El Bassam, N. (2013). *Renewable Energy sources and their applications.* Jodhpur, India: Agrobois International.

BP. (2017). *BP energy outlook 2017 BP energy outlook.* London, UK: British Petroleum.

CCSA. (2012). *CCS roadmap.* Retrieved from https://www.gov.uk/government/uploads/system/uploads/attachment_data/file/48317/4899-the-ccs-roadmap.pdf.

CE. (2017). *Working mine methane.* Retrieved from https://www.clarke-energy.com/coal-gas/working-mine-methane/.

CT. (2018). *Energy management: A comprehensive guide to controlling energy use*. Retrieved from https://www.carbontrust.com/resources/guides/energy-efficiency/energy-management/.

EIA. (February 13, 2015). *Shale gas and tight oil are commercially produced in just four countries*. Retrieved from https://www.eia.gov/todayinenergy/detail.php?id=19991.

EPA. (2012). *Guidelines for energy management*. New York, USA: US Environmental Protection Agency.

Ford, H., & Crowther, S. (1922). *My life and work*. New York, USA: Doubleday Page & Company.

Guney, M. S., & Tepe, Y. (2017). Classification and assessment of energy storage systems. *Renewable and Sustainable Energy Reviews, 75*, 1187–1197.

Harvey, C., Beardsmore, G., Moeck, I., & Rüter, H. (2016). *Geothermal exploration - global strategies and applications*. Bochum, Germany: IGA Academy Books.

IEA. (2013). *Energy efficiency market report 2013*. Paris, France: International Energy Agency.

IEA. (2016). *Energy efficiency market report 2016*. Paris, France: International Energy Agency.

IEA. (2017). *Energy efficiency 2017*. Retrieved from https://www.iea.org/publications/freepublications/publication/Energy_Efficiency_2017.pdf.

IHA. (2016). *A brief history of hydropower*. Retrieved from https://www.hydropower.org/a-brief-history-of-hydropower.

IRENA. (2014). Global bioenergy supply and demand projections. In D. S. A. D. G. Shunichi Nakada (Ed.), *A working paper for REmap 2030*. Abu Dhabi, UAE: International Renewable Energy Agency.

Kavalov, B., & Peteves, S. (2007). The future of coal DG JRC Institute for Energy. *European Communities, Luxemburg*.

Martinás, K. (2005). Energy in physics and in economy. *Interdisciplinary Description of Complex Systems, 3*(2), 44–58.

Matek, B. (2016). *Annual U.S. & global geothermal power production report 2016*. USA: Geothermal Energy Association.

Sattista. (2017). *Global marine energy capacity from 2009 to 2016 (in megawatts)*. Retrieved from https://www.statista.com/statistics/476267/global-capacity-of-marine-energy/.

Schwab, K. (2016). *The fourth industrial revolution: What it means and how to respond*. Retrieved from https://www.weforum.org/agenda/2016/01/the-fourth-industrial-revolution-what-it-means-and-how-to-respond/.

Sentryo. (2017). *The 4 industrial revolutions*. Retrieved from https://www.sentryo.net/the-4-industrial-revolutions/.

Shaw, G. (May 7, 2013). *Liquefied petroleum gas (LPG), liquefied natural gas (LNG) and compressed natural gas (CNG)*. Retrieved from http://www.envocare.co.uk/lpg_lng_cng.htm.

UNDP. (2000). *Energy and the challenge of sustainability*. New York, USA: United Nations Development Program.

VDI. (2007). *Energy management - terms and definitions*. Berlin, Germany.

WCA. (2009). *The coal resources: A comprehensive overview of coal*. UK: World Coal Association.

WEC. (2016). *World Energy Resources 2016*. UK: Retrieved from London. https://www.worldenergy.org/wp-content/uploads/2016/10/World-Energy-Resources-Full-report-2016.10.03.pdf.

WNA. (2017). The Nuclear Fuel Report. *Global Scenarios for Demand and Supply Availability*, 2017–2035. https://www.world-nuclear.org/our-association/publications/publications-for-sale/nuclear-fuel-report.aspx.

WWEA. (2018). *WWEA bulletin.* Retrieved from http://www.wwindea.org/information-2/publications/.

Yang, M., & Yu, X. (2015). *Energy efficiency: Benefits for environment and society.* Berlin, Germany: Springer.

Further reading

Mauthner, F., Weiss, W., & Spörk-Dür, M. (2016). Solar Heat Worldwide: Markets and Contribution to the Energy Supply 2014. Retrieved from Austria.

Chapter 2

Modern energy conversion technologies

M.M. Islam, M. Hasanuzzaman, A.K. Pandey, N.A. Rahim
Higher Institution Centre of Excellence (HICoE), UM Power Energy Dedicated Advanced Centre (UMPEDAC), Level 4, Wisma R&D, University of Malaya, Jalan Pantai Baharu, Kuala Lumpur, Malaysia

2.1 Introduction

One of the central challenges for mankind is to transform the present energy structure into a sustainable one. Energy-conversion methods and energy-conversion efficiency play a crucial part to accomplish this goal. The sole purpose of energy conversion is to convert the inoperative form of energy available in nature into useful forms, such as work. Developments in energy-conversion processes and devices like the steam engine or nuclear power throughout time mark breakthroughs of eras in the history of human civilization. Over the centuries, a range of systems and technologies has been developed for efficient conversion of energy wherein fossil fuel energy conversion was the main motive; in contrast, current energy-conversion technologies cover renewable energies.

2.2 Fossil fuel energy conversion

Energy in fossil fuels (oil, gas, and coal), originally captured from the sun through photosynthesis, is nonrenewable in nature. Fossil fuels are all hydrocarbons and upon burning, carbon dioxide (CO_2) and water (H_2O) molecules are formed with the release of a huge amount of heat. This heat is utilized to produce rotation (in turbines) or translation (in heat engines), which is finally converted into electrical energy through generators. The major fossil fuel conversion technologies are briefly discussed in the subsequent sections.

2.2.1 Internal combustion engines

The internal combustion (IC) engine is a class of heat engine wherein the chemical energy of fuel is transformed into shaft work. It is so named because

Energy for Sustainable Development. https://doi.org/10.1016/B978-0-12-814645-3.00002-X

combustion occurs inside a combustion chamber that is an integral part of the working fluid flow circuit. The two basic components of an IC engine are a stationary cylinder and a motile piston, the piston being pushed down by growing combustion gases inside the cylinder, which in succession revolves the crankshaft and by way of a gear system in the power train drives the vehicle.

Combustion in IC engines may be intermittent or continuous. IC engines with intermittent combustion are spark ignition (SI) gasoline and compression ignition (CI) diesel engines. Most are four-stroke engines including four distinctive processes, viz., intake, compression (and combustion), power, and exhaust stroke. The distinction between SI and CI engine consists in the method of igniting the fuel. In an SI engine, the fuel is first blended with air and then drafted into the cylinder during the intake process, while in a CI engine, only air is inducted into the engine and compressed, after which diesel fuel is injected into the hot compressed air at a suitable measured rate resulting in ignition (Heywood, 2018).

IC engines with continuous combustion include gas turbines, jet engines, and most rocket engines. Although typically IC engines are fed with fossil fuels, the use of alternative fuels like biodiesel in CI engines and bioethanol or methanol in SI engines is growing day by day. Recently, hydrogen as a fuel for IC engine has also been in the experimental application stage.

2.2.2 Steam turbines

A steam turbine is the single prime technology that has been employed globally in base load power plants to provide continuous electricity supply throughout the year. About 85% of all electricity in the USA in 2014 was generated by steam turbines, and the global scenario is almost identical (Manushin, 2011). Steam turbine units with as high as a 1500 MW capacity are commercially available, whereas most gas turbine units do not exceed 750 MW capacity, which is why most base load plants comprise steam turbines.

In steam turbines (Fig. 2.1), heated and compressed steam (generated by a boiler or from natural origin like geothermal wells) is allowed to expand in the turbine blade cascades, through which potential energy is transformed into kinetic energy and drives a shaft (BLAIR, 2017). Steam turbines work on the Rankine cycle comprising four processes, viz., an isentropic compression of saturated liquid (i.e., pumping water from condenser to boiler), heat addition at constant pressure (to produce steam inside the boiler), isentropic expansion of the steam (through the turbine), and constant pressure heat rejection from the steam (in the condenser to obtain saturated liquid). Hence, a steam turbine power plant consists of the major components, viz., boiler (steam generator), turbine, condenser, and feed pump along with some auxiliary systems: lube oil system, steam condensate system, etc. (STP, 2019).

Steam turbines offer better thermal efficiencies than reciprocating engines and gas turbines and offer greater reliability when sustained high power output

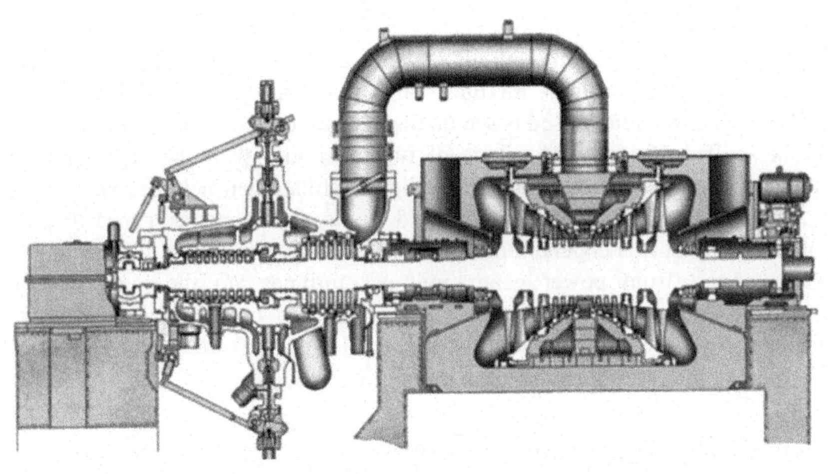

FIGURE 2.1 Steam turbine cutaway view Breeze, P. (2014). Power Generation Technologies, 2nd Edition, Imprint: Newnes, Elsevier.

is required. However, longer start-up time and efficiency drop at part-load operation are some of the few shortcomings of this device.

2.2.3 Gas turbines

Gas turbines (Fig. 2.2) convert the potential energy of heated and compressed gas into kinetic energy as a result of its expansion in several stages of turbine blading that eventually rotate the turbine shaft to realize mechanical work. This continuous combustion IC engine work on the Joule-Brayton cycle (J-B)

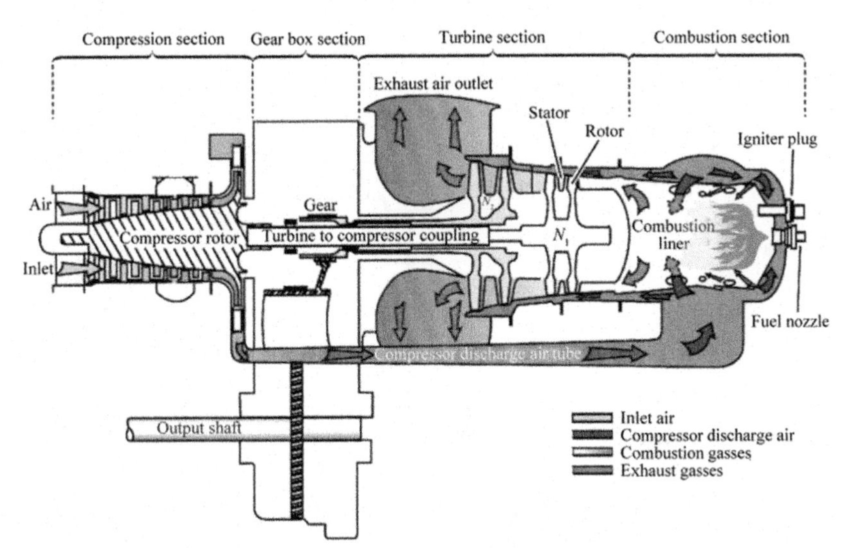

FIGURE 2.2 Gas turbine. *From Nkoi, B., Pilidis, P., & Nikolaidis, T. (2013). Performance assessment of simple and modified cycle turboshaft gas turbines. Propulsion and Power Research, 2(2), 96–106.*

(also known as the Brayton cycle) and comprises a compressor, a combustor or burner, and a turbine. Although air is the standard working fluid for the J-B cycle, inert gases or their mixtures or the combustion products of fossil fuels are also frequently used based on the application (STP, 2019). In fact, gas turbines offer the good advantage of fuel flexibility with the capability of adopting almost any flammable gas or light distillate petroleum products.

Gas turbines are primarily employed in large-scale power generation in solitary setup or in a cogenerating installation along with steam turbine power plants. Apart from power generation, gas turbines are the most widely employed propulsion systems in modern aircraft. In the aviation sector, different designs of gas turbines (popularly called engine) are available such as turbofan engine, turboprop engine, turbojet engine, etc.

2.2.4 Combined-cycle power plants

Combined-cycle power plants (Fig. 2.3) are compound gas turbine–steam turbine systems wherein the extreme hot exhaust from a gas turbine is employed to run a boiler, and the steam thus produced is fed into a steam turbine to generate power. These plants can deliver high power output at efficiencies as high as 50%–60% with low emissions and produce 50% more electricity than a simple-cycle plant consuming the same amount of fuel (Ramireddy, 2012). Combined cycle power plants may be either single-shaft, wherein both of the gas turbine and steam turbine are connected to the same generator in a tandem arrangement, or multishaft, with each gas turbine and steam turbine driving a separate generator.

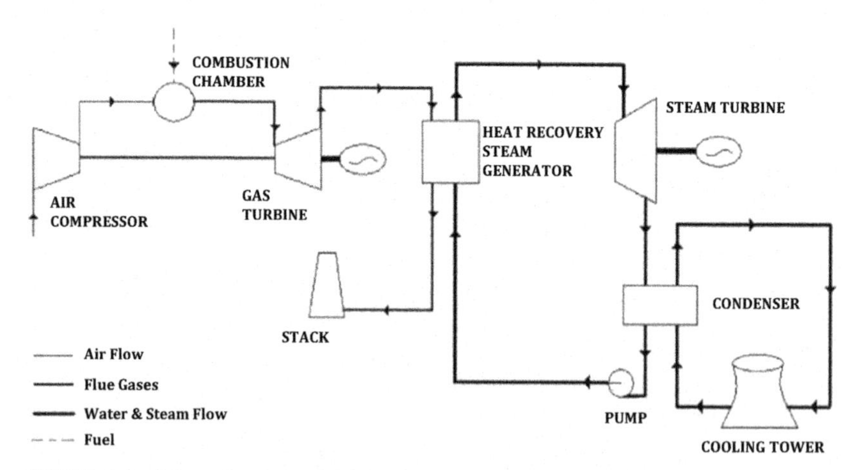

FIGURE 2.3 Schematic of combined-cycle power generation. *Dev, N., Kachhwaha, S. S., & Attri, R. (2014). Development of reliability index for combined cycle power plant using graph theoretic approach. Ain Shams Engineering Journal, 5(1), 193–203.*

2.2.5 Advanced combustion technologies

2.2.5.1 Clean coal technology

Coal is the world's most plenteous and widely disseminated fossil fuel source and fulfills about 27% of global primary energy needs. Almost 70% of world steel production and 38% of global electricity generation is dependent on coal feedstock (USC, 2019). However, it is the dirtiest of all fossil fuels, and emissions from its combustion contribute to global warming, create acid rain, and pollute water. It has been estimated that every year more than 14 billion tonnes of CO_2 are emitted from coal burning (USC, 2019).

Clean coal technology, as shown in Fig. 2.4, is a combination of technologies developed in attempts to diminish the negative environmental impact of coal energy generation. "Clean coal" connotes supercritical coal-fired plants without carbon capture and storage (CCS), since CO_2 emissions are less than for conventional plants (IEA, 2008). Coal cleaning removes primarily pyritic sulfur along with other impurities such as Pb, As, Hg, Ni, Sb, Se, and Cr and thereby improves the heating value of the coal.

2.2.5.2 Pulverized coal combustion

Coal-based power plants with conventional stoke firing suffer from some shortcomings such as the inability to handle load fluctuations due to limited combustion capacity, difficulties in removing large quantities of ash, and interference of the formed ash in the combustion process. Pulverized coal combustion systems offer a viable solution to all these problems, wherein coal is reduced to a fine powder in grinding mills and projected into the combustion chamber through a hot primary air current. To ensure complete combustion, supplementary air (known as secondary air) is delivered separately to the combustion chamber. Globally, almost 97% of the coal power plants with sizes up to 1000 MW are run by pulverized coal. The efficiency of the pulverized fuel firing system mostly depends on the size of the powder, and average net efficiencies reach up to 35% (Nicol, 2013). Pulverized coal technology is classified (Table 2.1) based on the operating temperature and pressure, wherein

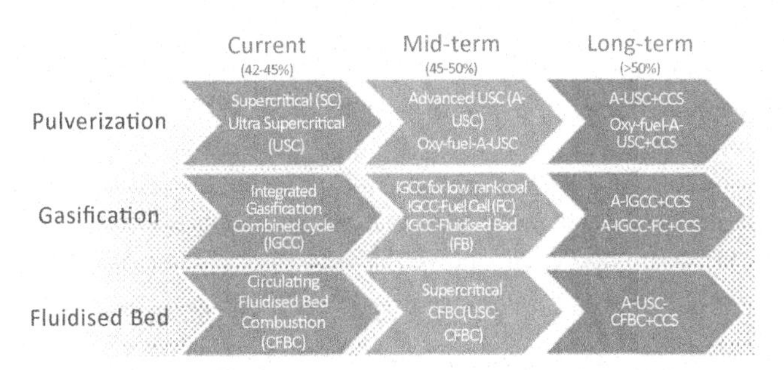

FIGURE 2.4 Clean coal technologies (Zhu, 2017).

TABLE 2.1 Classification of pulverized coal combustion plant with steam parameters, efficiency, etc. (Cocco, Reddy Karri, & Knowlton, 2014).

Technology	Super heating point	Efficiency (%)	Coal consumption (gm/kWh)
Subcritical	≤540°C <22.1 MPa	<35	≥380
Supercritical	540°C–580°C 22.1–25 MPa	35–40	380–340
Ultrasupercritical	580°C –620°C 22–25 MPa	40–45	340–320
Advanced ultrasupercritical	700°C–725°C 25–35 MPa	45–52	320–290

subcritical plants operate below the critical point of water, and the supercritical, ultrasupercritical, and advanced ultrasupercritical (AUSC) work beyond the critical point. The efficiency of these technologies increases (from <35% for subcritical to 52% for AUSC) with their operating temperatures and pressures.

2.2.5.3 Fluidized bed combustion

Fluidized bed combustion (FBC) is a specialized combustion process wherein solid particulates are suspended in upward jets of air in order to achieve more effective chemical reaction and heat transfer. This advanced technology is being used in the cracking of hydrocarbons, gasification of coal, roasting of ore, calcination of limestone, combustion of waste, etc. (NNFCC, 2009). One of the major benefits of FBC is that it facilitates the combustion of many unconventional fuels that would not be otherwise possible. In addition, it allows very low emission of NOx and SO_2.

2.2.5.4 Gasification technology

Gasification is the transformation of carbonaceous materials (either organic or fossil fuel) into CO, H_2, and CO_2 by means of reacting the materials with a regulated volume of oxygen and/or steam at high temperatures (>700°C) without burning them up. The resulting gas, which itself is a fuel, is called syngas or producer gas (Tucker, 2018). Gasifier technology is compliant with a wide range of fuels such as coals, petroleum coke, etc. and can also be employed in combined-cycle power plants—e.g., integrated gasification combined-cycle plants.

2.2.5.5 Carbon capture and storage

CCS is a cluster of technologies including accumulating, transporting, and then containing the carbon dioxide in order that it does not break out into the atmosphere and add to climate change. There are three key techniques in CCS: (1) converting coal into a clean-burning gas, (2) cleaning the power plant of exhaust gas using chemicals, and (3) burning the coal with pure oxygen of high concentration resulting in almost pure CO_2 exhaust. After trapping CO_2, it is liquefied, transported, and buried either in suitable geological formations, deep underground saline aquifers, or abandoned oil fields. Referring to the last option, CO_2 is sometimes pumped into oil fields to force out the residual pockets of oil (known as enhanced oil recovery) that would otherwise be tough to extract (Tanaka & Hasanuzzaman, 2018; Walker, 1980).

2.3 State-of-the-art energy conversion technologies

2.3.1 Stirling engine

The Stirling engine (Fig. 2.5) is a closed-cycle regenerative heat engine with permanently sealed gas, wherein the heat source is generated external to the engine. The engine is designed such that the gas is compressed in the cooler part and expanded in the hotter part ensuing a net conversion of heat into work (Andersson et al., 2015). Although air is the traditional working fluid for Stirling engines, hydrogen and helium are also employed for improved performance due to their better thermal properties.

Combustion occurs continuously in Stirling engines, which ensures clean burning. Moreover, the efficiency of this engine is not susceptible to the fuel

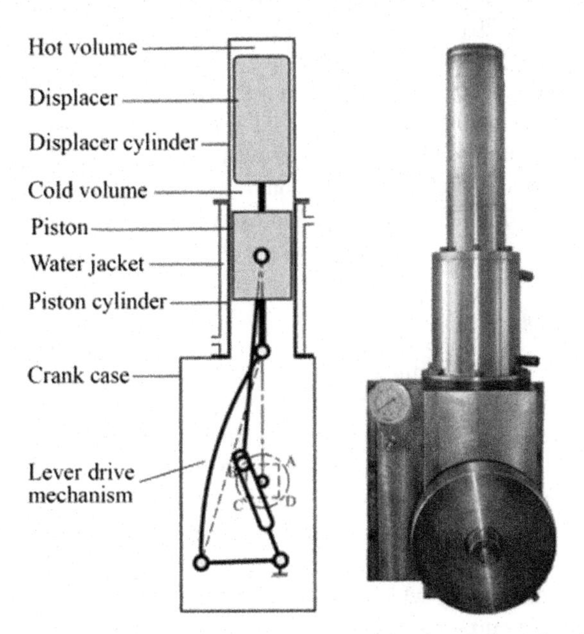

FIGURE 2.5 Stirling engine schematic *(Source: Wang, K., Sanders, S. R., Dubey, S., Choo, F. H., & Duan, F., 2016).*

type or quality. That is why the Stirling engine is more effectual than SI 2engines for hybrid electric automotive applications. In addition, the Stirling engine is attuned with both conventional and renewable energy sources, making it a future solution to fossil fuel depletion and climate change problems.

2.3.2 Nuclear power

Nuclear power has been the most concentrated and high-quality energy available to date. The energy contained in 1 kilogram of uranium, if it were all released, would produce energy equivalent to that produced from firing 3000 tonnes of coal. A nuclear reactor discharges nuclear energy as heat that is employed to generate steam; the steam is fed to a conventional steam turbine to generate electricity (DEC. Duke Energy Corporation, 2013). Although most nuclear power plants employ nuclear fission (splitting an atom into two), nuclear fusion (combining atoms into one) has good potential to be a benign way to produce power (Juan, Douglas, David, Paul, & Darla, 2012).

Different types of nuclear reactors have been developed to facilitate controlled release of nuclear energy, and many are still under development. Some major reactor technologies are pressurized water reactor, boiling water reactor, pressurized heavy water reactor, gas-cooled reactor, light water-cooled graphite-moderated reactor, fast breeder reactor, high-temperature gas-cooled reactor, and very-high-temperature reactor (EPA. U.S. Environmental Protection Agency, 2015).

Nuclear power plants are designed to run unceasingly, and it has been observed that nuclear plants run at their full capacity more than 90% of the time without any maintenance, making it the most reliable energy source. However, the problem with these power plants is the radioactive waste they produce in the course of power generation. It has been estimated that the USA alone produces 2000 metric tonnes of radioactive waste every year (Un-Noor, Padmanaban, Mihet-Popa, Mollah, Hossain, 2017).

2.3.3 Fuel cells

Fuel cells are electrochemical cells that produce electricity directly by extracting the chemical energy of a fuel and an oxidant eluding the wasteful multistep processes of heat engines, thereby enhancing efficiency and reducing emissions. Fuel cells operate silently without vibration and hence are suitable for onsite usage. Due to the simplicity of their construction and operation, fuel cells are apposite for distributed and portable power generation. Moreover, these devices can cope quickly with varying load conditions without compromising efficiency, which makes them suited for part-load operation. Above all, for the same quantity of energy-conversion, fuel cells produce less

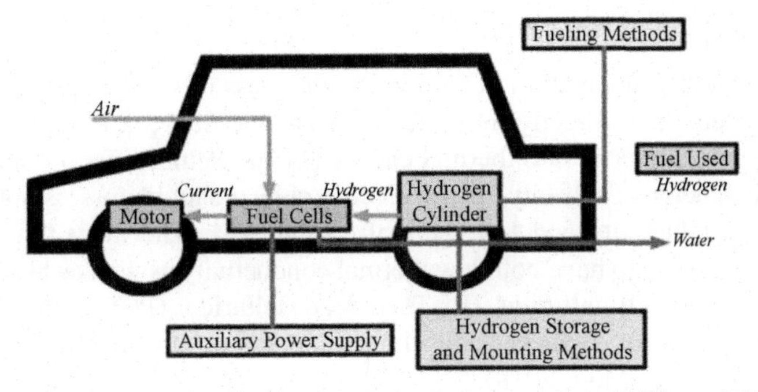

FIGURE 2.6 Basic components of a fuel cell electric vehicle (Williams, 2004).

emissions because of high conversion efficiency. Fuel cells are frequently used in space applications, zero-emission vehicles (Fig. 2.6), etc.

Currently, six major fuel cell technologies are at different stages of development and commercialization: alkaline, phosphoric acid, polymer electrolyte membrane, molten carbonate, solid oxide, and direct methanol fuel cells (Chang, 1982).

2.3.4 Thermionic power conversion

Thermionic energy conversion (TEC) is the direct generation of electricity by thermionic electron emission, wherein intensely hot electron vapor works as the working fluid in cycle (Baksht et al., 1978). When refractory metals like cesium or tungsten are heated to temperatures as high as 2000 K, electrons are vaporized from their surface (known as the emitter) to create a plasma and then cross the small interelectrode gap to a colder surface (known as the collector) at around 1000 K where they condense, creating a voltage that impels the current through the load and returns it to the emitter. TECs are high-current low-voltage devices, and depending on the emitter temperature, usually 25–50 A/cm^2 at a voltage of 1.0–2.0 V is achievable through this process. Like other cyclic heat engines, the maximum efficiency of thermionic generators is limited by Carnot's law and typically efficiencies of 5%–20% are practicable (Geller et al., 1996). The magnitude of current and voltage depends on electrode surface properties, and the rate of transport of electrons from emitter to collector is determined by plasma properties. Cesium vapor is employed in most converters because it can be easily ionized of all stable elements. Although TEC is a high-temperature device, research shows that at lower operating temperatures, significant enhancements in converter performance are possible if oxygen is added to the cesium vapor (Ramalingam & Young, 1993).

The TEC can produce power in the range of 5 kWe to 5 MWe, and electrical power in this range can work for telecommunication satellites, navigation, propulsion, and terrestrial exploration missions (Leighton et al., 2015).

2.3.5 Thermoelectric generators

Thermoelectric generators (TEGs) refer to any class of solid-state device that either converts heat directly into electricity (by the Seebeck effect) or transforms electrical energy into thermal energy (by the Peltier effect) (Goldsmid, 2017). To generate electricity, thermoelectric objects should possess both good electrical conductivity and low thermal conductivity. The main three semiconductors known to have both low thermal conductivity as well as high power factor are bismuth telluride (Bi_2Te_3), lead telluride (PbTe), and silicon germanium (SiGe). However, all three compounds are highly expensive (Teffah, Zhang, & Mou, 2018).

A thermoelectric module (Fig. 2.7) comprises two different thermoelectrics, with n-type and p-type semiconductor substances connected at their ends. As the two ends are maintained at different temperatures, DC electricity will generate in the circuit, the magnitude being directly proportional to the temperature difference. TEG modules are subjected to high thermal stresses and strains for extended periods of time, which also creates mechanical fatigue. Thus, selection of junctions and materials should ensure resistance to these harsh conditions.

The efficiency of a thermoelectric module is greatly dependent on its design. Usually thermoelectric materials should be arrayed thermally in parallel but electrically in series. Despite the fact that they have no moving parts, thus eliminating friction losses, TEGs are still less efficient. However, TEGs can utilize the waste heat of conventional power plants to produce supplementary electrical power. Automotive thermoelectric generators are another practical application of this technology. Moreover, radioisotope thermoelectric generators are used in space probes.

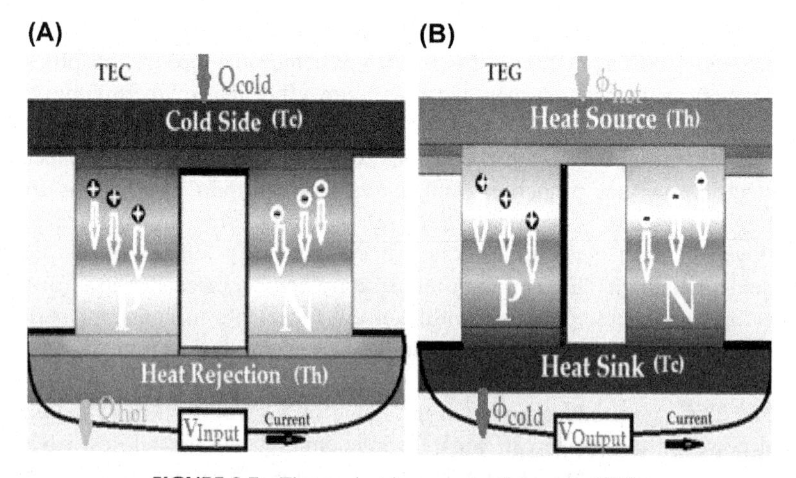

FIGURE 2.7 Thermoelectric generator (Messerle, 1995).

2.3.6 Magneto-hydrodynamic power generation

A magneto-hydrodynamic (MHD) generator, similar to a regular generator, generates electricity by means of revolving a conductor over a magnetic field, where instead of copper a hot conductive plasma is used as the moving conductor (Ajith Krishnan & Jinshah, 2013). In fact, MHD generators are one of the initiatives of scientists to eliminate mechanical systems working in between thermal and electrical energy conversion, with a view to reduce the loss associated with thermodynamic conversion. That is, MHD generators are employed to generate electricity directly from heat without any intermediate moving parts. An electrically conductive gas under high pressure is flown over a magnetic field (stemmed from the Hall effect), then the gas is expanded to run a turbine. Exclusion of moving parts in MHD generators facilitates running at higher temperatures (as high as 3000 K) than conventional heat engines, which could deliver efficiencies near 90% (Klinghoffer & Castaldi, 2013).

2.3.7 Waste-to-energy conversion

Waste-to-energy (WtE) is a specially designed energy generation facility that employs household waste as fuel. It has been estimated that from one tonne of municipal solid waste, 750 kWh of power and 22.68 kg of metal can successfully be recovered. WtE conversion facilitates reduce the volume of waste by almost 90% and save space in the landfill. One hundred cubic yards of waste can be burnt to only 10 cubic yards of ash with a significant amount of electricity supply, equivalent to 13,000 kWh. Currently, the best technology for WtE conversion is incineration in combined heat and power (CHP) plants, wherein solid waste volume can be decreased ninefold (Overend, Milne, & Mudge, 2012).

2.3.7.1 Thermochemical conversion

Incineration with mixed waste input is the most popular thermochemical conversion technology applied through CHP plants.

Cocombustion of solid waste with a supplementary fuel (usually coal or biomass) helps to regulate thermal properties.

Refuse-derived fuel technology can be employed for energy production either by monocombustion or cocombustion with municipal solid waste or coal (Angelidaki, Karakashev, Batstone, Plugge, & Stams, 2011).

Thermal gasification transforms carbonaceous materials into energy-rich gases, which can be a worthy alternative to burning out the wastes.

2.3.7.2 Biochemical conversion

Several methods for biodecomposition of wastes into energy-rich fuels are as follows:

Bioethanol is manufactured by chemical treatment of some organic wastes through hydrolysis (via enzymatic treatment), fermentation (using

microorganisms), and distillation (Angenent, Karim, Al-Dahhan, Wrenn, & Domíguez-Espinosa, 2004).
Dark fermentation and photofermentation are practices to mend organic materials into hydrogen without light or under light in turn through the activity of different bacteria. These technologies are applied mainly for wastewater treatment (Zaman, 2009).
Biogas is the raw methane gas (unfiltered and unpurified) from anaerobic digestion processes (Angenent et al., 2004).
Landfill gas is a mixture of gases (40%–60% methane) created by the action of microorganisms within landfill sites (Min, Cheng, & Logan, 2005).
Microbial fuel cells produce electricity by means of the conversion of the chemical energy content of organic substances by means of the catalytic reaction of microorganisms and bacteria (Hamad, Agll, Hamad, & Sheffield, 2014).

2.3.7.3 Chemical conversion (esterification)

Esterification is the process of producing ester through the reaction of an alcohol and an acid. Various biofuels can be obtained from wastes through esterification (IRENA, 2018).

2.4 Renewable energy conversion systems

Renewable energy still lacks a unanimous definition, with definitions split by type of energy sources included and in sustainability criterion adopted. The International Renewable Energy Agency has a statutory definition, ratified by 108 members (107 states and the European Union) as of February 2013 (IEA, 2002): "Renewable energy includes all forms of energy produced from renewable sources in a sustainable manner, including bioenergy, geothermal energy, hydropower, ocean energy, solar energy and wind energy."

The International Energy Agency (IEA) defines renewable energy resources as those "derived from natural processes and replenished at a faster rate than they are consumed." The IEA definition of renewable energy includes the following sources: "electricity and heat derived from solar, wind, ocean, hydropower, biomass, geothermal resources, and biofuels and hydrogen derived from renewable resources."

2.4.1 Solar energy conversion systems

Basically, there are two types of solar energy-conversion systems—solar thermal energy-conversion systems and solar electrical or photovoltaic (PV) systems. In recent years, there have been developed hybrid photovoltaic thermal systems that provides both electricity and heat form the same module (Fayaz, Rahim, Hasanuzzaman, Nasrin, Rivai, 2019a, Fayaz, Rahim, Hasanuzzaman, Rivai, Nasrin 2019b; Gary Cook, 1995).

2.4.1.1 Solar photovoltaics

PV or solar cells are semiconductor devices that produce electricity when exposed to sunlight through photovoltaic effect. Although PV modules (electrically connected PV cells) constitute the heart of a PV power system, a number of other components are required to conduct, control, convert, distribute, and store the energy produced by the PV array (an assembly of electrically connected PV modules). Depending on the functional and operational requirements, the DC—AC power inverter, battery bank, battery controller, and auxiliary energy sources are integrated with PV modules (or arrays). In addition, overcurrent, surge protection, disconnect devices, and other power processing equipment, known as the balance of system, are also included in a PV power system (Fig. 2.8) (SPS, 2019).

Primarily crystalline silicon (c-Si) cells (both mono- and multicrystalline) are used in PV panels, though recently the use of amorphous silicon (a-Si) cells has been growing steadily. The most favorable feature of a-Si is that only 1% of the material (silicon) is required to produce an a-Si cell as compared to that required to produce a c-Si cell of the same size. Apart from silicon cells, nowadays thin-film (cadmium telluride; copper-indium diselenide; copper-indium-gallium disulphide, and CIGS) are becoming commercially worthwhile due to their lower production costs. Recently, organic solar cells such as dye-sensitized solar cells are becoming popular.

2.4.1.2 Solar thermal technologies

Low-temperature solar thermal systems includes unglazed flat-plate solar collectors and evacuated tube collectors and are employed in temperature requirements not exceeding 180°C. These collectors employ the "greenhouse

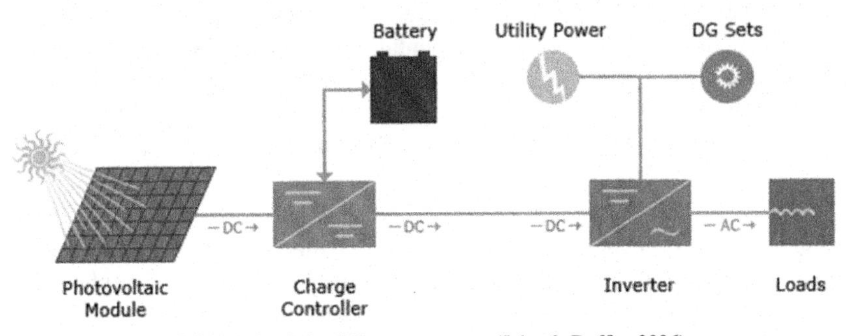

FIGURE 2.8 Solar PV power system (John & Duffie, 2006).

effect" to arrest the high-energy short-wavelength incident radiation to transform into long wave and gradually heat up an absorber plate. Applications include water and space heating, swimming pool heating, etc. (Fayaz et al., 2019a, 2019b; IRENA, 2013; Nahar, Hasanuzzaman, Rahim, & Parvin, 2019).

Concentrating solar power systems focus incident sunrays along a line or to a point. Line focus collectors track the sun along a line in a single axis and raise temperatures from 100 to 550°C. On the other hand, point focus collectors concentrate incident sun rays at a point receiver, creating temperatures as high as 800°C or more (Heimsath, Platzer, Heß, Krüger, & Eck, 2009; Islam, Hasanuzzaman, & Rahim, 2015; Kumar, Hasanuzzaman, & Rahim, 2019) shows different concentrating technologies with their temperature ranges and applications.

2.4.2 Bioenergy technologies

Bioenergy is the energy derived from materials of biological origin that is not fossilized. Raw woody biomass and animal wastes meet the world's major share of primary energy need; however, these can be metamorphosed into various gaseous or liquid fuels.

2.4.2.1 Biomass for power and heat

Thermochemical (combustion, pyrolysis, gasification, and liquefaction) or biochemical (digestion and fermentation) gasification of cellulosic biomasses such as wood chips, pellets or wood powder, or agricultural wastes like straw or husks transform these things into a flammable gas (Anna Schnürer, 2010).

2.4.2.2 Biogas

Biogas is a blend of gases produced by the decomposition of organic matter in the absence of oxygen. It is used for heating or cooking, running IC engines, or as synthetic gas to yield higher-quality fuels or chemicals. To produce biogas, biomass is first heated in lack of air to break down the solid mass into gas, the gas is then cleaned, and lastly filtered to remove unwanted chemicals. Anaerobic digestion, a form of fermentation, is another way to transform organic matter into biogas. Anaerobic biogas, which is comparable to landfill gas, comprises 60% methane and 40% carbon dioxide (Uriarte, 2010). Biogas is used in power generation systems such as combined-cycle power plants.

2.4.2.3 Biofuels

Liquid biofuels cover pure plant oil (also called straight vegetable oil), biodiesel, and bioethanol (Omidvarborna, Kumar, & Kim, 2014). Ethanol is principally produced from sugar, maize, and other starch-rich crops. Biodiesels are vegetable oil- or animal fat-based diesel replacements comprising long-chain alkyl esters, usually produced by reacting lipid with alcohols

(Ackermann, 2005). Biodiesel can be used in regular diesel engines in pure form or may be blended with petro diesel in any proportion; thus 100% pure biodiesel is denoted B100, 20% biodiesel blended with 80% petro diesel is denoted B20, and so on. On the other hand, bioethanol is ethyl alcohol (C_2H_5OH) produced from the fermentation of glucose, sucrose, etc. from plant sources (sugarcane, corn, sugar beet, etc.) by engaging microorganisms like yeasts or bacteria.

2.4.3 Wind energy conversion systems

Wind energy is the kinetic energy of blowing air as it flows due to atmospheric pressure gradients, and wind turbines are the contrivance to convert this kinetic energy into electricity. Wind power that could be harvested is proportional to the rotor diameter and cube of the wind speed; hence, as wind speed is doubled, wind power potential upsurges by eight times. There are two basic configurations of wind turbines, namely vertical axis wind turbines and horizontal axis wind turbines (HAWTs); a vast majority of installations are HAWTs with either two or three vanes. Large turbines are grouped together to form a wind power plant, which feeds power to the electrical transmission system (TSSWT, 2012).

2.4.4 Ocean or marine energy technology

The waves and tides of the oceans involve enormous energy potential that can be harvested for practical applications. Moreover, temperature difference between cold deep waters and hot surface waters and the salinity gradient at river mouths can also be encountered for power generation (Ruud Kempener, 2014). The established and potential marine energy technologies are as follows.

2.4.4.1 Ocean thermal energy conversion

Solar energy reserved as thermal energy in the upper layers of the oceans constitute an enormous renewable source estimated to be as high as 10^{13} W (SERI, 1989). Ocean thermal energy-conversion (OTEC) technology harnesses energy by using the temperature gradient between the warmer ocean surface water and the cooler deep water (Hamedi & Sadeghzadeh, 2017). It has been observed that even a temperature difference of 20°C can provide operative energy (Barstow, Mollison, & Cruz, 2008).

OTEC plants are of three types: open (Claude) cycle, closed (Anderson) cycle, and hybrid cycle. Claude cycle makes use of the tropical ocean's warm surface water as a heat transfer fluid (HTF) wherein warm water is allowed to expand rapidly in a partially evacuated chamber causing a phase change to steam, and as the steam continues to expand it drives a low pressure steam turbine. At the turbine exit, by using cold seawater, the vapor is condensed to freshwater containing no salt, which can be a source of drinking water for nearby townships. A 1-MW OTEC can produce around 4500 m^3 of freshwater per day, adequate to provide for a 20,000 populace. The Anderson cycle employs ammonia, propane, or Freon as the working HTF, which is confined in an entirely closed system including the turbine. Hot surface water is employed to vaporize the fluid through a heat exchanger, and the vapor produced thereby is expanded over the vanes of a turbine which in turn drives a generator. At the turbine exit, the vapor is condensed with cool deep ocean-water and pumped back to the evaporator. In hybrid cycles, instead of discharging the hot water of a closed-cycle OTEC to the ocean, it is fed to an open-cycle OTEC system to augment power output.

2.4.4.2 Wave energy

Wave energy is a blend of both kinetic and gravitational potential energy of ocean waves created as a result of the wind blowing over the surface. Wave resources are extremely variable, as availability is strongly dependent on wave height, and during storms wave power can reach highs of 200 kW/m, but are insecure for maneuvering. The average yearly incident wave power at the northeastern Pacific and Atlantic is about 50 kW/m, while near Cape Horn in South America this figure may reach as high as 100 kW/m (Goswami & Kreith, 2017).

Electricity can be generated by harnessing wave energy by means of an oscillating water column that comprises an air compartment in contact with the ocean; as the water column oscillates, it makes the air to flow in and out, thereby turning a turbine. Well turbines with symmetrical airfoil blades counterbalanced by flywheels (to transform the oscillation motion due to waves into linear motion) are used to deliver power.

2.4.4.3 Tidal power

Tides contain huge amounts of potential energy that can generate power during the gravity-driven inflow or outflow (or both) of seawater through a turbo-generator (Dye, 2012). A dam is erected to isolate the tidal basin from the sea and create a difference in water level between them. During high tide, water flows from the sea into the basin through the turbine and generates power; on the other hand, during low tide, as water flows out from tidal basin to the sea it rotates the turbine in opposite direction and produces power, too.

2.4.5 Geothermal power generation

Geothermal energy is heat energy generated and stored in the Earth's crust and originates from the radioactive decay of materials in the original formation of the planet (DiPippo, 2012). While the global average geothermal gradient is 30°C/km, the regional gradient can exceed 90°C/km, which is adequate for power generation, and this high-temperature energy is available only in a few places. Geothermal energy can be ranked as high temperature (>180°C), intermediate temperature (101 to 180°C,) and low temperature (30 to 100°C) energy.

There are four categories of geothermal power plants: dry steam plants, flash power plants, binary geothermal plants, and flash/binary combined cycle. Dry steam noncondensing cycle is the basic and inexpensive way to generate electricity, wherein steam from the geothermal well is flown over a turbine and spent to the atmosphere. However, the most common type of geothermal plant is the flash steam power plant, which mainly addresses water-dominated reservoirs with temperatures greater than 180°C. The hot water is pumped up to the surface until its pressure drops and water starts boiling, creating a two-phase mixture of water and steam. Thus, steam is separated from the water, going to the turbine, with the cooled water piped back to source. Binary-cycle power plants use the heat of the hot geothermal fluid (74–177°C) to boil a working fluid (typically an organic compound of low boiling point), the geothermal fluid and the working fluid are confined to separate closed loop fluids to generate steam through a heat exchanger and then run a turbine. Apart from the above configurations, there are binary-cycle plants, which have three configurations, viz., brine bottoming binary, spent steam bottoming binary, and hybrid system (WorldAtlas, 2019). Although geothermal power provides about 0.4% of global power generation, the growth rate is stable. With 22 power plants with a combined installed capacity of 1.52 GW, the Geysers Complex, in the Mayacamas Mountains, San Francisco, California, USA, is the biggest geothermal field in the world (Breeze, 2018).

2.4.6 Hydropower generation

Hydroelectric power (briefly called hydropower) is derived from water in motion. Water stored at a higher level is allowed to flow down through conducting lines and flow over the vanes of a turbine installed downstream of the flow; as the water turns the turbine, it runs the generator to generate electricity. In order to generate a reliable supply of water with significant potential energy, dams are essential; the dam creates a head from which water flows through a pipe (called penstock) from the headwater reservoir through the turbine to the tailwater reservoir. Apart from conventional hydro dams, electricity can be produced from run-of-the-river by harnessing the kinetic energy in rivers or streams, pumped-storage reservoirs, etc. In addition, microhydropower and picohydropower plants are attracting substantial attention from researchers (IHA, 2018). Depending on the head available, hydroelectric power plants employ mainly three categories of water turbines: a Pelton wheel for very high heads (>1500 m), Francis turbines for medium head (~750 m), and Kaplan turbines (<100 m).

In 2017, the installed capacity of world hydropower was 1267 GW with an estimated generation of 4185 TWh (AFP, 2015). The Three Gorges Dam over the river Yangtze in Hubei, China, is the world's largest hydropower plant with a generation capacity of 22.5 GW, while Itaipu in Brazil/Paraguay over the river Parana has a capacity of 14.0 GW (AFP, 2015).

References

Ackermann, T. (2005). *Wind power in power systems*. West Sussex, England: John Wiley & Sons, Ltd.

AFP. (2015). China's Three Gorges dam 'breaks world hydropower record'. In *MailOnline*. UK: Associated Newspapers Ltd.

Ajith Krishnan, R., & Jinshah, B. S. (2013). Magnetohydrodynamic power generation. *International Journal of Scientific and Research Publications, 3*(6).

Andersson, N., Eriksson, L., & Nilsson, M. (2015). Numerical simulation of stirling engines using an unsteady quasi-one-dimensional approach. *ASME Journal of Fluids Engineering, 137*(5). https://doi.org/10.1115/1.4029396. 2015, 051104-051104-9.

Angelidaki, I., Karakashev, D., Batstone, D. J., Plugge, C. M., & Stams, A. J. (2011). Biomethanation and its potential. In *Methods in enzymology* (pp. 327–351). Elsevier.

Angenent, L. T., Karim, K., Al-Dahhan, M. H., Wrenn, B. A., & Domíguez-Espinosa, R. (2004). Production of bioenergy and biochemicals from industrial and agricultural wastewater. *Trends in Biotechnology, 22*(9), 477–485.

Anna Schnürer, Å. J. (2010). Microbiological handbook for biogas plants. In *Swedish gas centre report 207*. Sweden: Swedish Waste Management.

Baksht, F., Dyvzhev, G., Martsinovskiy, A., Moyzhes, B. Y., Dikus, G. Y., Sonin, E., & Yuryev, V. (1978). *Thermionic converters and low-temperature plasma*. NASA STI/Recon Technical Report N.

Barstow, S.,M. G., Mollison, D., & Cruz, J. (2008). The wave energy resource. In C. J. (Ed.), *Green energy and technology*. New York, USA: Springer Berlin Heidelberg.

BLAIR, T. H. (2017). *Energy production systems engineering*. New Jersy, USA: IEEE Press, John Wiely & Sons Inc.

Breeze, P. (2018). *Hydropower*. Massachusetts, USA: Academic Press Elsevier.

Breeze, P. (2014). Power Generation Technologies. In *Imprint: Newnes* (2nd Edition). Elsevier.

Chang, S. S. (1982). Fundamentals handbook of electrical and computer engineering. In *Circuits fields and electronics* (Vol. 1). New York: Wiley-Interscience, 716 pp., 1982.

Cocco, R., Reddy Karri, S. B., & Knowlton, T. (2014). *Introduction to fludization*. Available from: www.aiche.org/sites/default/files/cep/20141121.pdf.

Dev, N., Kachhwaha, S. S., & Attri, R. (2014). Development of reliability index for combined cycle power plant using graph theoretic approach. *Ain Shams Engineering Journal, 5*(1), 193–203.

DEC. Duke Energy Corporation. (2013). *Fission vs. Fusion – what's the difference?* Web.

DiPippo, R. (2012). *Geothermal power plants: Principles, applications, case studies and environmental impact* (3rd ed.). Waltham, USA: Butterworth-Heinemann.

Dye, S. T. (2012). Geoneutrions and the radioactive power of the Earth. *Reviews of Geophysics, 50*(3), 1–19.

EPA. U.S. Environmental Protection Agency. (2015). *Solid waste generation*. Web.

Fayaz, H., Rahim, N. A., Hasanuzzaman, M., Nasrin, R., & Rivai, A. (2019a). Numerical and experimental investigation of the effect of operating conditions on performance of PVT and PVT-PCM. *Renewable Energy, 143*, 827–841.

Fayaz, H., Rahim, N. A., Hasanuzzaman, M., Rivai, A., & Nasrin, R. (2019b). Numerical and outdoor real time experimental investigation of performance of PCM based PVT system. *Solar Energy*, 135–150.

Gary Cook, L. B. (1995). *Rick adcock photovoltaic fundamentals*. Washington DC, USA: National Renewable Energy Laboratory and DOE National Laboratory.

Gasturbine. (2019). *Gasturbine*. https://www.mhps.com/products/gasturbines/lineup/h100/.

Geller, C., Murray, C., Riley, D., Desplat, J., Hansen, L., Hatch, G., … Rasor, N. (1996). *High efficiency thermionics (HET-IV) and converter advancement (CAP) programs*. Final reports. Sunnyvale, CA: Rasor Associates, Inc. (United States); Bettis Atomic Power ….

Goldsmid, J. (2017). *The physics of thermoelectric energy conversion*. San Rafael CA, USA: Morgan & Claypool Publication (IOP Concise Physics).

Goswami, D. Y., & Kreith, F. (2017). *Energy conversion* (2nd ed.). Florida, USA: CRC Press.

Hamad, T. A., Agll, A. A., Hamad, Y. M., & Sheffield, J. W. (2014). Solid waste as renewable source of energy: Current and future possibility in Libya. *Case Studies in Thermal Engineering, 4*, 144–152.

Hamedi, A.-S., & Sadeghzadeh, S. (2017). Conceptual design of a 5 MW OTEC power plant in the Oman Sea. *Journal of Marine Engineering and Technology, 16*(2), 94–102.

Heimsath, A., Platzer, W., Heß, S., Krüger, D., & Eck, M. (2009). *Concentrating solar collectors for process heat and electricity generation*. FVEE-AEE Topics.

Heywood, J. B. (2018). *Internal combustion engine fundamentals* (2nd ed.). USA: McGraw-Hill Education.

IEA. (2002). *World energy outlook 2002*. Paris, France: International Energy Agency.

IEA. (2008). *Clean coal technologies: Accelerating commercial and policy drivers for deployment*. Paris Cedex, France: International Energy Agency (IEA).

IHA. (2018). In R. Taylor (Ed.), *Hydropower status report: Sector trends and insightts*. London, UK: International Hydropower Association.

IRENA. (2018). *Renewable energy statistics 2018*. Abu Dhabi: International Renewable Energy Agency.

Islam, M. K., Hasanuzzaman, M., & Rahim, N. A. (2015). Modelling and analysis of the effect of different parameters on a parabolic-trough concentrating solar system. *RSC Advances, 5*(46), 36540–36546.

John, A., & Duffie, W. A. B. (2006). *Solar engineering of thermal processess* (3rd ed.). New Jersy, USA: John Wiley and Sons.

Juan, S. G., Douglas, J. G., David, G. N., Paul, V. P., & Darla, J. M. (2012). *Fundamentals of nuclear power*. USA: State Utility Forecasting Group.

Kehlhofer, R. (1997). *Combined-cycles gas and steam turbine power plants*. Tusla,Oklahoma, USA: PennWell Publishing Co.

Klinghoffer, N., & Castaldi, M. (2013). *Waste to energy conversion technology*. Swaston, UK: Woodhead Publishing, Elsevier.

Kumar, L., Hasanuzzaman, M., & Rahim, N. A. (2019). Global advancement of solar thermal energy technologies for industrial process heat and its future prospects: A review. *Energy Conversion and Management, 195*, 885–908.

Leighton, E., Sissom, G., & Ralph, S. (2015). *Thermionic power converter*, 21/5]; Available from: www.britannica.com/technology/thermionic-power-converter.

Manushin, E. A. (2011). *Steam turbine*. Thermopedia.

Messerle, H. K. (1995). *Magnetohydrodynamic electrical power generation*. New Jersy, USA: John Wiley & Sons Inc.

Min, B., Cheng, S., & Logan, B. E. (2005). Electricity generation using membrane and salt bridge microbial fuel cells. *Water Research, 39*(9), 1675–1686.

Nahar, A., Hasanuzzaman, M., Rahim, N. A., & Parvin, S. (2019). Numerical investigation on the effect of different parameters in enhancing heat transfer performance of photovoltaic thermal systems. *Renewable Energy, 132*, 284–295.

Nicol, K. (2013). *Status of advanced ultra-supercritical pulverised coal technology*. London, UK: IEA Clean Coal Centre.

Nkoi, B., Pilidis, P., & Nikolaidis, T. (2013). Performance assessment of simple and modified cycle turboshaft gas turbines. *Propulsion and Power Research, 2*(2), 96–106.

NNFCC. (2009). *Review of technologies for gasification of biomass and wastes*. UK: National Non-Food Crops Center.

Omidvarborna, H., Kumar, A., & Kim, D.-S. (2014). Characterization of particulate matter emitted from transit buses fueled with B20 in idle modes. *Journal of Environmental Chemical Engineering, 2*(4), 2335–2342.

Overend, R. P., Milne, T., & Mudge, L. (2012). *Fundamentals of thermochemical biomass conversion*. London, UK: Springer.

Patterson, W. C. (1986). *Nuclear power* (2nd ed.). Harmondsworth, Middlesex, England: Penguin Books Ltd.

Ramalingam, M. L., & Young, T. J. (1993). The power of therionic energy conversion. *Mechanical Engineering, 115*(9), 78–83.

Ramireddy, V. (2012). *An overview of combined cycle power plant*. Available from: https://electrical-engineering-portal.com/an-overview-of-combined-cycle-power-plant.

Ruud Kempener, F. N. (2014). *Ocean thermal energy conversion:technology brief*. Abu Dhabi, UAE: International Renewable Energy Agency (IRENA).

SERI. (1989). *Ocean thermal energy conversion: An overview*. Golden CO, USA: Solar Energy Reserach Institute.

SPS. (2019). *Solar photovoltaic systems*. http://www.synergyenviron.com/resources/solar-photovoltaic-systems.

STP. (2019). *Steam turbine plant*. https://www.indiamart.com/neha-engineeringworks/products.html#steam-turbine-plant.

Tanaka, Y., & Hasanuzzaman, M. (2018). A review of global current techniques and evaluation methods of photocatalytic CO_2 reduction. In *IET conference publications*.

Teffah, K., Zhang, Y., & Mou, X.-l. (2018). Modeling and experimentation of new thermoelectric cooler–thermoelectric generator module. *Energies, 11*(3), 576.

TSSWT. (2012). *Toward super-size wind turbines: Bigger wind turbines do make greener electricity*. https://www.constantinealexander.net/2012/06/toward-super-size-wind-turbines-bigger-wind-turbines-do-make-greener-electricity.html.

Tucker, O. (2018). *Carbon capture and storage*. Bristol, United Kingdom: IOP Publishing Ltd.

Un-Noor, F., Padmanaban, S., Mihet-Popa, L., Mollah, M. N., & Hossain, E. (2017). A comprehensive study of key electric vehicle (EV) components, technologies, challenges, impacts, and future direction of development. *Energies, 10*(8), 1217.

Uriarte, F. A. (2010). *Biofuels from plant oils*. Jakarta, Indonesia: ASEAN Foundation.

USC. (2019). *All About Coal: Coal's impacts include air and water pollution, worker deaths, and climate change*. [cited 2019]; Available from: https://www.ucsusa.org/clean-energy/coal-impacts.

Walker, G. (1980). *Stirling engines*. Oxford, UK: Clarenden Press.

Wang, K., Sanders, S. R., Dubey, S., Choo, F. H., & Duan, F. (2016). Stirling cycle engines for recovering low and moderate temperature heat: A review. *Renewable and Sustainable Energy Reviews, 62*, 89–108.

Williams, M. C. (2004). *Fuel cell handbook* (7th ed.). Morgantown, West Virginia, USA: National Energy Technology Laboratory (NETL).

WorldAtlas. (May 17, 2019). *Largest geothermal power plants in the world.* Available from: https://www.worldatlas.com/articles/largest-geothermal-power-plants-in-the-world.html.

Zaman, A. U. (2009). Life cycle environmental assessment of municipal solid waste to energy technologies. *Global Journal of Environmental Research, 3*(3), 155–163.

Zhu, Q. (2017). *Power generation from coal using supercritical CO_2 cycle.* London, UK: IEA Clean Coal Center.

Further reading

Cavanagh, J. E., Clarke, J. H., & Price, R. (1993). *Ocean energy systems.* In T. B. Johanson, H. Kelley, A. K. N. Reddy, & R. H. Williams (Eds.). Renewable energy: Sources for fuel and electricity.

Chapin, D., Kiffer, S., & Nestell, J. (2004). *The very high temperature reactor: A technical summary.* Alexandria VA, USA.

EIA, U. (March 2015). *Electricity net generation.*

Harpster, G. R. S. J. W. (May 15, 2019). *Thermoelectric power generator.* Available from, www.britannica.com/technology/thermoelectric-power-generator 3.5.

HEPP, Hydroelectric power plant working | types of hydroelectric power plants, 2019 http://electricalacademia.com/renewable-energy/hydroelectric-power-plant-working-types-hydro electric-power-plants/ Press, Washington DC, USA

IRENA. (2013). Concentrating solar power. In *IRENA-IEA-ETSAP technology briefs.* Abu Dhabi, UAE: International Renewable Energy Agency.

McKendry, P. (2002). Energy production from biomass (part 2): Conversion technologies. *Bioresource Technology, 83*(1), 47–54.

WNA. (2018). *Clean coal' technologies, carbon capture & sequestration.*

Chapter 3

Energy demand

M. Hasanuzzaman[a], M.A. Islam[a], N.A. Rahim[a], Yuan Yanping[b]
[a]*Higher Institution Centre of Excellence (HICoE), UM Power Energy Dedicated Advanced Centre (UMPEDAC), Level 4, Wisma R&D, University of Malaya, Jalan Pantai Baharu, Kuala Lumpur, Malaysia;* [b]*School of Mechanical Engineering, Southwest Jiaotong University, Chengdu, China*

3.1 Introduction to energy demand

The term "energy demand" normally refers to any type of energy required to fulfill individual or sectoral energy needs. Individual energy demand relates to the individual energy requirements for fulfillment of different purposes: cooking, heating, cooling, etc. Sectoral energy demand relates to the energy requirements of different sectors such as industrial, residential, and transportation. Energy demand can correspond to: (1) ***primary energy demand***—this is the amount of energy required by a country, or (2) ***final energy demand***—this is the amount of energy supplied to consumers. Energy demand narrates the relationship between the price and quantity of energy in the form of electricity or fuel. It normally demonstrates what amount of energy will be bought at a given cost and how price changes will influence that amount. The whole energy system of a country is derived according to energy demand. The overall worldwide energy demand depends on not only total energy use but also location, available energy resources, resource types and properties, characteristics of end-use technology, etc.

3.2 Demand classification

Demand can be classified as follows:

- ➢ Elastic demand
- ➢ Inelastic demand
- ➢ Unit elastic demand

3.2.1 Elastic demand

Elastic demand is demand where price or other factors have a big impact on consumer needs (i.e., demand). Demand is elastic when the quantity demanded changes significantly in response to a change in price. The consumer will buy a

Energy for Sustainable Development. https://doi.org/10.1016/B978-0-12-814645-3.00003-1

lot more if the price of the commodity goes down just a little. And if prices increase just a little bit, consumers will reduce their purchasing and wait for normal prices to return. Consumers will purchase goods or services having elastic demand by making many comparisons, because consumers are not desperate to buy specific goods or services because many alternatives are available. So in the case of elastic demand, the quantity purchased (i.e., quantity demanded) has an inverse relationship with price.

3.2.1.1 Elastic demand formulation

The demand for goods or services is considered elastic when the percentage change in the quantity demanded exceeds the percentage change of price. So elastic demand can be formulated as follows (KENTON, 2019; Khan, 2019):

$$\text{Demand Elastic} = \frac{\nabla D}{\nabla P} > 1$$

where

∇D is the percentage change in the quantity demanded
∇P is the percentage change in price

For example, the quantity demanded of a good increases by 10% when its price falls by 5%. The ratio is 0.10/0.05 = 2.00. So demand for the product is elastic.

3.2.2 Inelastic demand

The demand for a good or service is considered inelastic when the percentage change in the quantity demanded is less than the percentage change in price. So inelastic demand can be formulated as follows (KENTON, 2019; Khan, 2019):

$$\text{Demand Inelastic} = \frac{\nabla D}{\nabla P} < 1$$

where

∇D is the percentage change in quantity demanded
∇P is the percentage change in price

For example, if the quantity demanded of a good rises by 2% when the price falls by 5%, the ratio is 0.02/0.05 = 0.40, which is less than 1. So, the demand for the good is considered inelastic.

3.2.3 Unit elastic demand

When the percentage change in quantity demanded for a good or service is equal to the percentage change in price, its demand is considered unit elastic, and the ratio between the percentage change in quantity demanded and the percentage change in price equals 1. Unit elastic demand can be formulated as follows (KENTON, 2019; Khan, 2019):

$$\text{Demand\_Unit\_Elastic} = \frac{\nabla D}{\nabla P} = 1$$

where

∇D is the percentage change in the quantity demanded
∇P is the percentage change in price

For example, if the quantity demanded increased 3% in response to a price drop of 3%, the ratio would be 0.03/.03 = 1.

3.2.4 Demand curve

The demand curve is a visual portrayal of the number of units of a commodity or service will be purchased for each of a range of conceivable prices. It represents the relationships between the various quantities and prices of a good or service. Fig. 3.1 shows a typical demand curve, where the price is on the vertical (y) axis and the demand is on the horizontal (x) axis. This is a conventional relationship between demand and price. Usually, the demand curve is downward-sloping (i.e., its slope is negative) because the number of units demanded increases as price falls and vice versa.

Higher price results in lower demand, and lower price results in higher demand.

3.3 Energy demand analysis in different sectors

Energy demand can be classified into many sectors by end user. According to the International Energy Agency (IEA, 2017), energy demand sector classifications are as follows:

➢ residential
➢ commercial and public services

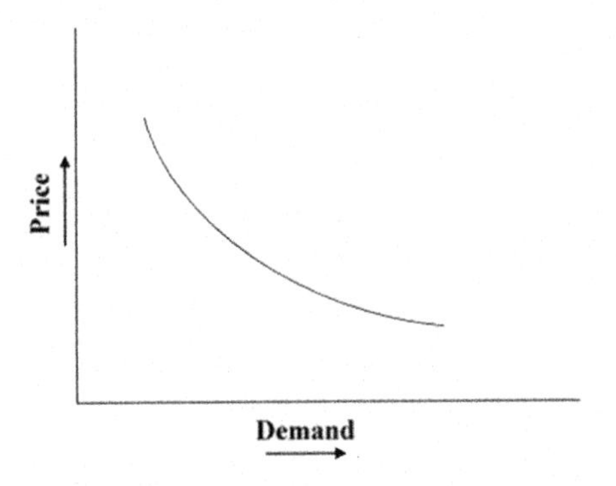

FIGURE 3.1 Basic demand curve.

- ➢ industrial
- ➢ transport
- ➢ nonenergy
- ➢ other#

Note, #: other sector includes agriculture, fishing, and nonspecified other (IEA, 2017).

According to the Malaysian Energy Commission (MESH, 2015), major common classifications by sector are as follows:

- ➢ residential
- ➢ commercial
- ➢ industrial
- ➢ transport
- ➢ agricultural
- ➢ fishery
- ➢ nonenergy

From the above classifications, sectoral energy demand analysis can be grouped as follows:

- ➢ building (residential and commercial)
- ➢ industrial
- ➢ transport

3.4 Building sector energy demand analysis

The building sector consumes a significant portion of energy demand; according to the US Energy Information Association, the building sector accounts for 20.1% of total worldwide energy consumption (EIA, 2016). Energy use intensity (EUI) depends on many factors (Ouf & Issa, 2017).

Energy consumption of a building's system mainly depends on its

- structural condition
- building materials
- weather
- operating condition

3.4.1 Structural condition

Structural condition and building materials are important factors influencing energy consumption. Energy consumption varies based on

- building structure/type
    - apartment
    - condominium

- flat
- bungalow
- duplex
- townhouse

3.4.2 Building materials

Structural building materials are important factors in energy consumption. Energy consumption varies based on

- material types
- thermal properties
- dimensions
- operating behaviors

Energy consumption in the building sector depends on the equipment used by households, including energy used for heating, cooling, lighting, cooking, and other electrical equipment:

- type of building
- number of inhabitants
- usage of the building
- amount of equipment
- power rating
- operating hours
- equipment efficiency

Energy consumption in buildings also depends on weather conditions:

- building location (hot or cold region)
- duration of each season
- temperature
- wind speed
- duration of day and night

The building sectors are commonly classified based on purpose and usage of buildings as follows:

- **Residential buildings**

Energy consumption in residential buildings varies significantly across provinces and countries. Different factors influence building energy consumption, such as building type and household characteristics, energy source availability and energy policies, homeowner income, and type of energy equipment used.

- Commercial buildings

Commercial buildings usually profit-seeking businesses or those that provide various services. Energy is consumed in commercial buildings by heating, lighting, cooling, computer and information system operating, etc. Commercial buildings include hospitals, offices, education institutions, police stations, warehouses, hotels and restaurants, shopping malls, retail stores, and places of worship. Some other on-building energy use in the commercial sector contributes to public services such as traffic lights and water and sewer systems.

3.4.3 Energy consumption of building systems

Worldwide building energy consumption is steadily increasing, and its contribution to global energy consumption ranges between 20% and 40% (Wei et al., 2018). Energy consumption of a building depends on different equipment usage behaviors of residents and occupants. To determine residential energy consumption, it is critical to understand resident or occupant behavior. Fig. 3.2 shows the relation between three concepts: energy consumption, building technology, and residential behavior. Resident behavior and actions can be divided into three categories (Zhao, McCoy, Du, Agee, & Lu, 2017):

- time-related usage,
- environment-related modes, and
- quantitatively described behavior

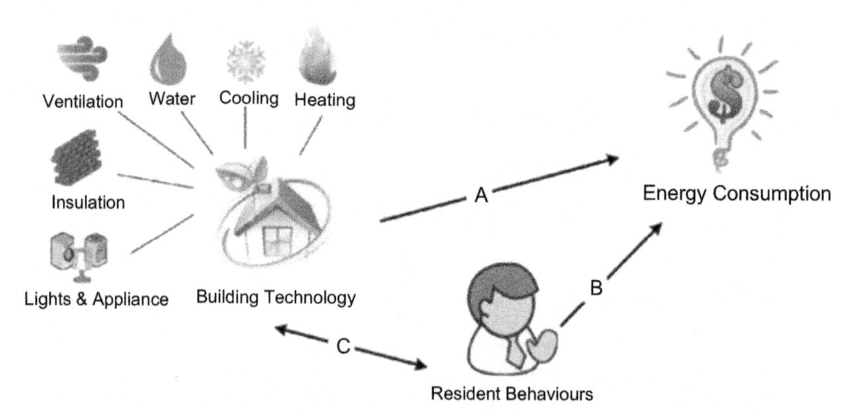

FIGURE 3.2 Relations between different main perceptions: building technology, resident behavior, and energy consumption (Zhao et al., 2017).

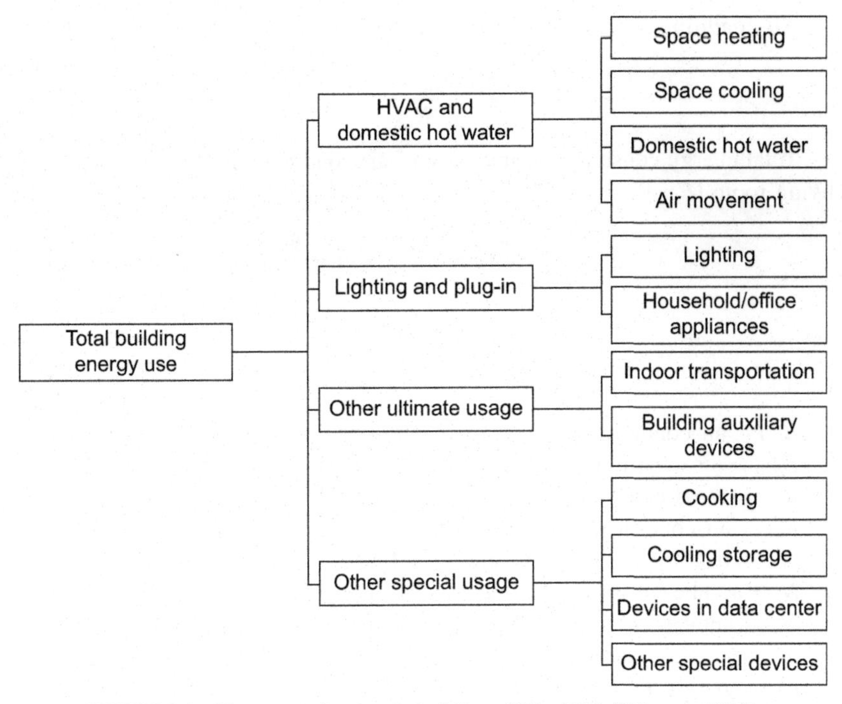

FIGURE 3.3 The usage of energy in buildings (ISO., 2013; Wei et al., 2018).

Different categories of equipment are responsible for building energy consumption. Fig. 3.3 shows different possible types of equipment for building energy consumption. Most buildings have some common equipment. Energy-consuming building equipment includes the following:

1. air conditioners
2. lights
3. lifts and escalators
4. household appliances

3.4.4 Air conditioner energy consumption

The air conditioner is a major energy-consuming component in a building system. A simplified energy consumption calculation formula for air conditioning system is as following equations 3.1-3.3 (Li, 2007; Yang, Liu, Huang, Min, & Zhong, 2015):

Air-conditioning energy consumption intensity (E_{acl}) in a living room of a house

$$E_{acl} = \frac{2.5 \times R_{op} \times \left(0.2181 \times T_{out}^2 - 7.1379 \times T_{out} + 60.406\right)}{R_{EE}} \tag{3.1}$$

Air condition energy consumption intensity (E_{acb}) in a bedroom of a house

$$E_{acb} = \frac{2.5 \times R_{op} \times \left(0.3128 \times T_{out}^2 - 712.238 \times T_{out} + 120.6\right)}{R_{EE}} \tag{3.2}$$

Total energy consumption intensity of air conditions in both bedroom and living room (E_{AC})

$$E_{AC} = \frac{(E_{acl} \times A_l \times N_l + E_{acb} \times A_b \times N_b)}{A_t} \tag{3.3}$$

where

E_{acl} = air-conditioning energy use intensity for the living room (kWh/m^2)
E_{acb} = air-conditioning energy use intensity for bedroom (kWh/m^2)
R_{EE} = rated energy efficiency ratio of the split air conditioner
R_{op} = air-conditioning operating ratio
T_{out} = average outdoor temperature (°C)
N_l = number of air conditioners in the living room of each house (set)
N_b = number of air conditioners in the bedroom of each house (set)
Nt = average quantity of air conditioners in each house (set)
A_l = area of the main living room (m^2)
A_b = area of main bedroom (m^2)
A_t = total area of each house (m^2)

3.4.5 Lighting energy consumption

Evaluating interior lighting energy—the following equations 3.4-3.5 applies to the calculation of the annual load (Brady & Abdellatif, 2017).

Annual energy use (kWh/year) = energy use for illumination (W_L) + parasitic energy (W_p)

$$W_L = \sum \{(P_t \times F_c) \times (t_d \times F_0 \times F_d) + (t_n \times F_0)\} \div 1000 \tag{3.4}$$

$$W_P = \sum (W_{pc} + W_{em}) \tag{3.5}$$

where

P_t = total power installed in room (W)
F_c = illuminance factor constant (assuming 1 as no constant illuminance control)
F_o = factor for occupancy dependent (assuming 0.9 as automatic control greater than 60% of connected load)
F_d = factor for daylight dependency (assuming 0.9 as photocell dimming with daylight sensing)

t_d = time of daylight (h)
W_{pc} = parasitic control annual energy consumption (5 kWh/m^2)
W_{em} = parasitic emergency annual energy consumption (1 kWh/m^2)

Lighting energy consumption can be assessed from the illumination area, residential lighting power density, and average time of use. The simplified calculation formula is as follows (Li, 2007; Yang et al., 2015):

$$E_L = P_L \times T_L \times R_L \tag{3.6}$$

where

E_L = intensity of lighting energy use (kWh/m^2)
P_L = average power density of light (kW)
T_L = average time of lighting in a year (h)
R_L = illumination area–to–floor area ratio

3.4.6 Lift and escalator

3.4.6.1 Lift energy calculation

Fig. 3.4 shows a typical lift system. The energy consumption of the lift and escalator system is affected by several factors (Sharif, 1996):

- different drives including
  - hydraulic lifts
  - two-speed lifts
  - eddy-current braking systems (DC injection braking)
  - variable-frequency drives
- mechanical design aspects
- various component efficiencies: gearbox, motor, etc.
- reduction of inertia: use of flywheels (and all other moving masses) reduces system efficiency
- gearing types: helical gears have higher efficiency than that of worm-wheel gears
- possibility of regeneration back into the mains: this depends on whether the system can return energy and whether the metering system can cope with reverse energy.
- power factor (especially important for Ward–Leonard systems)
- type of loading: level of usage and number of passengers

The maximum steady-state torque on the motor, T_{FL} (in N m), corresponding to the maximum number of passengers allowed is computed using the following equations 3.7-3.11 (Ahmed et al., 2014):

$$T_{FL} = (NP_{\max} \times MPP + M_{lift} - M_{ew}) \times g \times \frac{r}{GR} \tag{3.7}$$

where

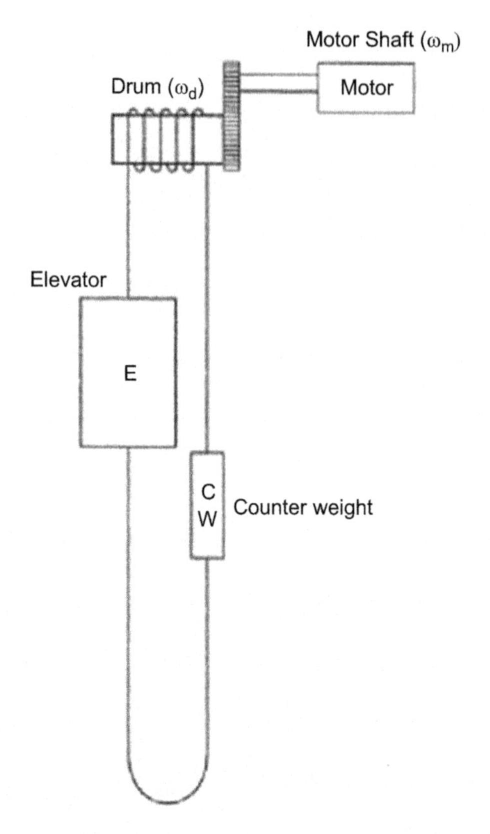

FIGURE 3.4 A lift system (the gear is not shown explicitly) (Ahmed, Iqbal, Sarwar, & Salam, 2014).

NP_{max} is the number of maximum passengers allowed in a lift
MPP is the average mass (kg) per passenger
M_{lift} is mass of the lift cabin in kg
M_{cw} is mass of the counterweight in kg
r is outer radius of the drum (m)
g is acceleration due to gravity
GR is the gear ratio

The shaft power, P_{FL} (horsepower—i.e., hp) required of the motor just to provide steady-state torque is:

$$P_{FL} = T_{FL} \times \frac{\omega_m}{746} = T_{FL} \times \left(\frac{v}{r}\right) \times \frac{GR}{746} \tag{3.8}$$

where motor angular velocity, ω_m, is equal to drum angular velocity, $\omega_d(=v/r)$, multiplied by the gear ratio. The motor inertia, J_m, in kg m^2 corresponding to P_{FL} (in hp) is estimated as

$$J_m \approx \frac{P_{FL}}{100} \tag{3.9}$$

From lift operating behavior, daily energy consumption by a single lift can be calculated as follows (Brady & Abdellatif, 2017):

$$E_L = (P_{motor} \times N_T \times t_s)/3600 + E_{standby} \tag{3.10}$$

where

E_L = daily energy consumption by lift (kWh/day)
P_{motor} = lift motor rated power (kW)
N_T = number of trips per day
t_s = time to travel between the main entrance floor and the highest-served floor from the instant the door has closed until the instant it starts to open (in s)
$E_{standby}$ = standby energy used by a single lift in 1 day (kWh)

The accuracy of the lift energy consumption estimation is largely affected by the parameter N_T, which needs very careful estimation or measurement. The annual energy consumption calculation by lift is as:

$$E_L(\text{kwh/year}) = E_L(\text{kWh/day}) \times N_{working-day} + E_{standby} \tag{3.11}$$

where

$N_{working\text{-}day}$ is the number of working days per year and
$E_{standby}$ is the standby energy used by a single lift in 1 year (kWh).

$$\text{Starts per day} = 2 * 240 + 2 * 240 + 8 * 40 = 1280 \text{ starts per day}$$

$$\text{E} = (45 \times 1280 \times 4)/(3600) = 64 \text{ kWh/day/lift}$$
$$\text{For all six lifts in the group, E} = 64 \times 6 = 384\text{kWh/day}$$
$$\text{The yearly consumption} = 384 \times 350(\text{working days per year})$$
$$= 134,400 kWh$$

Annual energy use to operate a single lift can be obtained from the method mentioned in CIBSE Guide D that originated on BS ISO/DIS 25745-1 (Brady & Abdellatif, 2017):

$$E_L = (N_T P_m t_h/4) + E_{standby} \tag{3.12}$$

where

E_L = energy used by a single lift in a year (kWh)
N_T = number of trips per year
P_m = motor drive power rating (kW)
t_h = time needed to travel between main entrance floor and topmost-served floor (from instant of door closing until instant of door opening) (h)

$E_{standby}$ = single lift standby energy consumption in a year (kWh)

However, the load and average travel distance are not always same throughout the lift operation time. So from the specific energy consumption (mWh/kg/m) of a lift, energy consumption can be calculated as follows (Tukia, 2014):

$$E_L = E_{Travel} + E_{standby} \tag{3.13}$$

$$E_{Travel} = E_{specific} \times T_{travel} \times v \times W_{Load} \tag{3.14}$$

$$E_L = (E_{specific} \times T_{travel} \times v \times W_{Load}) + E_{standby} \tag{3.15}$$

where

E_{Travel} is energy consumption by lift during travel
$E_{Standby}$ is energy consumption by lift during standby
$E_{Specific}$ is specific travel energy consumption by lift
T_{Travel} is the average travel time of the lift
V is velocity
W_{Load} is nominal load
$E_{Specific}$, specific travel energy consumption by lift, can be calculated by using following equation (Tukia, 2014)

$$E_{specific} = \frac{k_L \times E_{reference}}{W_{Load} \times 2 \times H} \tag{3.16}$$

where

k_L is the load factor
$E_{reference}$ is the measured energy consumption of reference trip including the full cycle up and down
H is the height of the lifting path

The load factor (k_L) can be derived using the following equation. According to the standard, the load factor for a 50% counterbalanced elevator is calculated as

$$k_L = 1 - (\%W_{Load} \times 0.0164) \text{ Range: } 0.97 - 0.74 \tag{3.17}$$

For traction elevators with 40% counterbalance, a different constant is used:

$$k_L = 1 - (\%W_{Load} \times 0.0192) \text{ Range: } 0.96 - 0.69 \tag{3.18}$$

The average load percentage ($\%W_{Load}$) depends on the usage categories and different rated load as follows (Barney & Lorente, 2013; Tukia, 2014): For $N_T \leq 500$, the average percent load can be considered 7.5% 4.5%, 3% and 2.0% for rated load $W_R \leq 800$ kg, 800 kg$< W_R <$1275 kg, 1276 kg$< W_R <$2000 kg and $W_{Load} > 2000$ kg respectively. For 500 $< N_T$

$\leq$1000, the average percent load can be considered 9.0%, 6.0%, 3.5% and 2.2% for rated load $W_R \leq 800$ kg, 800 kg$<W_R<$1275 kg, 1276 kg$<W_R<$2000 kg and $W_{Load} > 2000$ kg respectively. For 1000 $<N_T$ $\leq$2000, the average percent load can be considered 16.0%, 11.0%, 7.0% and 4.5% for rated load $W_R \leq 800$ kg, 800 kg$<W_R<$1275 kg, 1276 kg$<W_R<$2000 kg and $W_{Load} > 2000$ kg respectively.

The energy consumption by lift will vary according to lift run mode. So daily energy consumption by lift can be calculated by considering average travel distance per trip as a percentage of the full height of the installed lift (Barney & Lorente, 2013; Tukia, 2014):

$$E_L = N_T \times \%L_D \times k_L \times E_{reference} \tag{3.19}$$

where

N_T is the number of trips made per day
$\%L_D$ is the average travel distance per trip as a percentage of the full height of the installed lift
k_L is the load factor per trip
$E_{reference}$ is the measured energy consumption of reference trip including the full cycle up and down

The average travel distance in percentage $\%L_D$ *is generally taken as follows* (Barney & Lorente, 2013; Tukia, 2014). For $N_T \leq 1000$, the average travel distance ($\%L_D$) can be considered 100%, 68% and 44% for number of stops equal to 2, 3, and greater than 3 respectively. For $1000<N_T \leq 2000$, the average travel distance ($\%L_D$) can be considered 100%, 68% and 33% for number of stops equal to 2, 3, and greater than 3 respectively.

$$E_{spec} = \frac{0.7 \times 60 \text{ Wh}}{1000 \text{ kg} \times 2 \div 35 \text{ m}} = 0.6(\text{mWh/kg/m})$$

$$E_{standby,annual} = 365\left(\frac{\text{day}}{\text{year}}\right) \times 30(\text{W}) \times 21(\text{hour/day}) = 230\left(\frac{\text{kWh}}{\text{year}}\right)$$

$$E_{Travel,annual} = 365365\left(\frac{\text{day}}{\text{year}}\right) \times 0.6\left(\frac{\text{mWh}}{\text{kgm}}\right)$$

$$\times \underbrace{3(\text{hour/day}) \times 3600\left(\frac{\text{s}}{\text{hour}}\right) \times 2.0(\text{m/s})}_{traveleddistanceperday} \times 1000 \text{ kg}$$

$$= 4730\left(\frac{\text{kWh}}{\text{year}}\right)$$

$$E_{Total,annual} = E_{travel,annual} + E_{standby,annual} = 4960\left(\frac{\text{kWh}}{\text{year}}\right)$$

3.4.6.2 Escalator energy consumption

Fig. 3.5 simply demonstrates the different positions of an escalator. The energy consumption of and escalator comprises three individual contributions:

- power to pull passengers at an incline (including additional friction caused by passenger mass), $P_{variable}$
- power needed to overcome friction in mechanical parts of the escalator, P_{mech}
- power consumed by supporting operations and auxiliary systems, $P_{standby}$

During the passenger-carrying position of the escalator, friction comes from three sources:

- passenger mass
- weight of the steps, and
- chain weight and other parts.

The total power consumption (P_{total}) of an escalator can be determined using the following equations (Uimonen et al., 2016):

$$P_{total} = P_{\text{variable}} + P_{mech} \tag{3.20}$$

$$P_{\text{variable}} = P_{carryPeople} + P_{stepLoad} \tag{3.21}$$

$$P_{mech} = P_{step\&chain} + P_{handrail} \tag{3.22}$$

$$P_{total} = P_{carryPeople} + \underbrace{P_{stepLoad} + P_{step\&chain} + P_{handrail}}_{Friction} \tag{3.23}$$

where

P_{total} is total power needed to operate the escalator
$P_{variable}$ is power to passenger load
P_{mech} is power needed to overcome friction in the mechanical parts of the escalator
$P_{carryPeople}$ is power needed to carry people
$P_{stepLoad}$ is power needed to overcome friction as a result of passenger steps

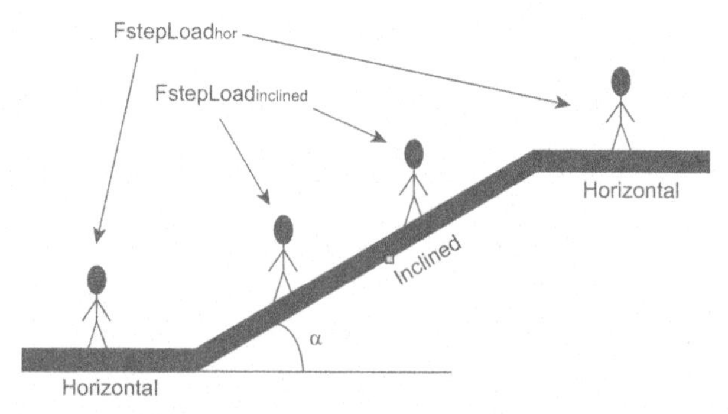

FIGURE 3.5 Different partitions of an escalator (Uimonen et al., 2016).

$P_{step\&chain}$ is power consumed due to the combined step and chain systems during a person's travel over the inclined part of the escalator
$P_{handrail}$ is power needed for the handrail
$P_{stepLoad}$ can be explained by the equation that follows. The first part of the equation is related to the power needed to overcome increased friction on the horizontal parts of the escalator, and the second part is related to the additional friction over the inclined portion of the escalator (Uimonen et al., 2016):

$$P_{stepLoad} = F_{stepLoad_{hor}} \times \upsilon \times \mu + F_{stepLoad_{incl}} \times \upsilon \times \mu \times \cos(\alpha) \tag{3.24}$$

where

$F_{stepLoad_{hor}}$ is the weight of people (vertical force due to gravity) on the horizontal steps,
$F_{stepLoad_{incl}}$ is the weight of people on the inclined steps,
v is the speed of the escalator steps,
μ is the friction coefficient, and
α is the angle of incline.

The power needed to carry the people, $P_{carryPeople}$, is expressed as follows (Uimonen et al., 2016):

$$P_{carryPeople} = F_{stepLoad_{incl}} \times \upsilon \times \mu \times \sin(\alpha)$$

$P_{step\&chain}$ can be expressed as follows (Uimonen et al., 2016):

$$P_{step\&chain} = \frac{N_{hor}}{N_{hor}+N} \times F_{step\&chain} \times \upsilon \times \mu + \frac{N_{hor}}{N_{hor}+N} \times F_{step\&chain} \times \upsilon \times \mu \times \cos(\alpha) \tag{3.25}$$

where

N_{hor} = total number of horizontal steps in both landings,
N = number of steps in inclined position, and
$F_{step\&chain}$ = combined weight of the step and chain systems including the returning truss while a person travels the inclined part of the escalator.

On the other hand, additional power is required for the handrail, $P_{handrail}$. Therefore, the total power of the escalator will be as follows (Uimonen et al., 2016):

$$P_{total} = \underbrace{(\upsilon \times (\mu \times (F_{stepLoad_{hor}} + F_{stepLoad_{incl}} \times \cos(\alpha)) + F_{stepLoad_{incl}} \times \sin(\alpha)))}_{variable} \\ \underbrace{+\left(F_{step\&chain} \times \upsilon \times \mu \times \left(\frac{N_{hor}}{N_{hor}+N} + \frac{N_{hor}}{N_{hor}+N} \times \cos(\alpha)\right) + P_{handrail}\right)}_{mechanical} \tag{3.26}$$

3.4.7 Energy consumption calculation of ventilation system

The energy consumption of a ventilation system consists of electricity used by its different components, such as electricity used by

- ventilation units,
- control devices, and
- pumps and other auxiliary devices.

The proficiency of energy utilized is evaluated based on the specific fan power of the ventilation framework at an ascertained wind stream rate. Fan-specific power is the proportion of the total power of the system to the wind stream rate [kW/(m^3/s)]. If the ventilation system is so much larger that the airflow rate is higher than 0.25 m^3/s, individual fan energy consumption must be calculated separately. The energy consumption of a ventilation system (E_v) is normally calculated using following formula (MEAC, 2014):

$$E_V = P_V \times \frac{\tau_d}{24} \times \frac{\tau_w}{7} \times t \tag{3.27}$$

where

P_v = power rate of the fan (kW)
τ_d = operating hours of fan at specific rate of airflow per day
τ_w = number of days of fan operating per week at of airflow-specific rate d;
t = time of 8760 h in regard to which the calculation is performed

The power consumption of ventilation system, P_v (W), can be calculated as follows (MEAC, 2014):

$$P_V = \frac{\nabla P \times V}{\eta_{ft}} \tag{3.28}$$

where

ΔP = arise of presser on the fan (Pa)
V = airflow rate of the fan (m^3/s);
η_{ft} = total efficiency of ventilation system

The total efficiency depends on different parameters including

- fan efficiency
- belt transmission efficiency
- motor efficiency, and
- efficiency of the control system of possible rotation speed.

The energy consumption of a small ventilation system whose airflow rate is lower than 0.25 m^3/s can be calculated as follows (MEAC, 2014):

$$E_{sv} = P_{sv} \times t_{svn} \times \chi_p \tag{3.29}$$

where

P_{sv} = power of the ventilation system at a specific airflow rate (kW),
t_{svn} = operating time of the ventilation unit in year at design air speed, and
χ_p = pressure drop factor of the ducts.

3.4.8 Household appliances energy consumption

Household appliances that consume energy that are commonly included:

- electric fan
- electric rice cooker
- electric cooker
- refrigerator
- TV
- computer
- washing machine
- microwave or electric oven
- kitchen ventilation
- water heater

Household appliance energy consumption depends on average power and average time of use. Energy consumption of household appliances can be calculated using the following formula (Yang et al., 2015):

$$E_{ha} = \frac{\sum N_{hai} \times P_{hai} \times T_{hai}}{A} \tag{3.30}$$

where

E_{ha} = household appliance energy intensity for individual house (kWh/m^2)
N_{ha} = number of household appliances in each house
P_{ha} = household appliance average power density (kW)
T_{ha} = average household appliance use time in a year (h)
A = average area of floor of individual house (m^2)

3.4.9 Evaluating energy use for small power appliances

This includes office equipment and other small power requirements for catering (microwave, toaster, kettle, coffee/vend, hand-dryer, and refrigerator). Modern small power equipment operates at working power and "sleep" condition. Annual energy use is determined from an assessment of power conditions and operating time (Brady & Abdellatif, 2017). The energy consumption per year of these appliances can be calculated as:

$$E_{sp} = (P_{sp} \times T_h) + E_{sleep} \times (8760 - T_h) \tag{3.31}$$

where

E_{sp} = annual energy consumption by small power appliances (kWh)
P_{sp} = average power consumption during operation

T_h = annual hours of operation
E_{sleep} = sleep mode energy consumption

Annual energy use to provide domestic hot water is found from the product of annual mass flow of water use and energy required to raise its temperature from 5°C (cold feedwater) to 65°C (storage temperature). Annual domestic hot water usage has been obtained from the CIBSE Guide G as follows (Brady & Abdellatif, 2017):

$$M_W = V_W \times D_W \times N_{day}) \tag{3.32}$$

where

M_W = mass of water (kg)
V_W = daily water consumption volume per person (L/person)
D_W = density of water (kg/L)
N_{day} = number of occupied days

Annual energy consumption of domestic hot water usage can be calculated as:

$$E_{hw} = M_W \times \Delta T \times C_w/3600 \tag{3.33}$$

where

E_{hw} = annual energy consumption (kWh/year)
M_W = mass of water (kg)
ΔT = temperature differential (k)
C_w = specific heat capacity of water (kJK/kg)

3.4.10 Building energy index

Building energy consumption varies according to building size area and usage of different equipment. For comparison purposes, an energy index called as building energy index (BEI) is used, which will determine whether the building over- or underconsumes energy. BEI is building energy consumption normalized by building size and is sometimes called the energy performance index, or EPI. So BEI is calculated simply as the total annual energy consumption of a building (kWh/year) divided by its total occupied area (m^2). Mathematically, BEI can be expressed as follows (BCIS, 2013):

$$BEI = \frac{EC_{year}}{A_{Floor}} \tag{3.34}$$

where

BEI is the building energy index in [kWh/year/m^2]
EC_{year} = annual total building energy consumption (kWh/year)
A_{Floor} = total occupied floor area of the building (m^2)

Total energy consumption is defined as the total amount of electricity or equivalent electricity consumed by the building per annum. If there is a renewable source for the building, then energy produced by the renewable energy source needs to be deducted from total energy consumption. BEI buildings provide a renewable energy source as follows (BCIS, 2013)

$$BEI = \frac{\sum EC_{year} - \sum EG_{RE}}{A_{Floor}} \tag{3.35}$$

where

EG_{RE} = energy generated from the renewable source per year (kWh/year)

3.4.11 Green building index

Building energy consumption is increasing as a consequence of increased building activity, especially in developing countries. The building sector is expected to consume as much energy as the industrial and transportation sectors consume. A grading system for buildings on the basis of energy efficiency for different facilities is a common practice. The green building index (GBI) is one of the building scoring systems for Malaysian buildings and was first announced in April 2009 as a combined endeavor of different institutes and organizations such as the Malaysian Institute of Architects and the Association of Consulting Engineers Malaysia, together with different building industries. In a GBI system, the building assessment is carried out on the basis of different criteria such as energy efficiency, sustainable site planning and management, quality of indoor environment, innovation, materials and resources, water efficiency, etc (Table 3.1).

3.5 Industrial sector

The industrial sector is one of the largest energy-consuming sectors and consumed about 54% of total world energy in 2012; its energy consumption is expected to grow 1.2% per year between 2012 and 2040 (EIA, 2016; Hasanuzzaman et al., 2012). In 2013, the Malaysian and Swedish industrial sectors consumed about 26.2% and 40%, respectively, of total energy usage by each country (MESH, 2015; SEA, 2018).

This sector can be divided by three different industry categories:

- energy-intensive manufacturing,
- nonmanufacturing, and
- non-energy-intensive manufacturing.

TABLE 3.1 Different green building rating system (Chua & Oh, 2011).

Building scoring tools	Full name	Country	Year	Criteria
BREEAM	Building Research Establishment Environmental Assessment Method	UK	1990	Management, health and well-being, energy, transport, water, materials, waste, land use, and ecology
LEED	Leadership in Energy and Environmental Design	USA	1998	Sustainable site, water efficiency, energy and atmosphere, materials and resources, indoor environmental quality, innovation, and design/ construction process
GS	Green Star	Australia	2003	Management, transport, land use and ecology, emissions, energy, materials, indoor environmental quality, and innovation
GM	Green mark	Singapore	2005	Energy efficiency, water efficiency, environmental protection, indoor environment quality, other green features, and innovation
GBI	Green building Index	Malaysia	2009	Energy efficiency, indoor environment quality, sustainable site and management, materials and resources, water efficiency, and innovation

3.5.1 Energy-intensive manufacturing (IEO, 2016)

Most industries are energy-intensive and major energy end-users of each country. The following industries are commonly considered energy-intensive:

- **food:** food, tobacco goods manufacturing, and beverages
- **chemicals:** organic chemicals, inorganic chemicals, agricultural chemicals, and resins, including chemical feedstocks
- **steel and iron:** steel and iron manufacturing including coke over

- **pulp- and paper-based:** printing and interrelated support activities and paper manufacturing
- **nonferrous metals:** mostly aluminum and other metals such as zinc, copper, and tin
- **refining:** coal products and petroleum refinery manufacturing including natural gas and coal used as feedstocks
- **nonmetallic minerals:** mainly cement and other nonmetallic minerals such as lime, glass, clay, and gypsum products

3.5.2 Non-energy-intensive manufacturing

- **other chemical industries:** pharmaceutical (botanical and medicinal), coatings and paint, detergents, adhesives, and other miscellaneous chemical yields
- **other industries:** the entire other industrial manufacturing including metal-based durables (machinery, fabricated metal items, electronic and computer products, electrical equipment, and transportation equipment)

3.5.3 Nonmanufacturing

- **forestry, agriculture, fishing:** forestry, agriculture, and fishing
- **construction:** industrial construction, civil and heavy engineering construction, building construction (commercial and residential), and specialty skill contractors
- **mining:** natural gas and oil extraction, coal mining, and mining of nonmetallic and metallic minerals

In the industrial sector, fuel mix and intensity consumed vary by country and region and depend on the mix of economic activity and technological development. Industrial sector energy is used for a variety of purposes—for instance, assembly and processing, cogeneration and steam, process cooling and heating, air conditioning for buildings, and heating. Energy consumption of the industrial sector also includes chemical feedstocks. Moreover, the feedstocks of natural gas are applied to generate agriculture chemicals. Both petroleum products (i.e., naphtha) and natural gas liquids are used in the manufacture of plastics and organic chemicals, among other uses.

Energy for the industrial sector is mainly consumed by the following equipment (Saidur et al., 2009):

- electric lights
- electric motors
- steam/hot water boilers
- air-conditioning systems
- refrigeration systems
- reciprocating air-compressor pumps
- ventilation and exhaust systems

- ❖ air cleaning equipment
- ❖ lifts/elevators/escalators
- ❖ conveyor systems
- ❖ cranes (overhead and gantry)
- ❖ workshop machines
- ❖ electric furnaces
- ❖ process heating/distinct heating

Fig. 3.6 shows energy consumption by various equipment types for different applications in typical industries in Malaysia.

3.5.3.1 Motor energy consumption

The electric motor converts power output to beneficial mechanical energy and consumes more electricity than other end uses. Electric motors are commonly schemed to run from 50% to 100% of rated load, and the maximum efficiency is typically about 75% of load. The efficiency of a motor tends to drop intensely below or near the 50% rated load (Table 3.2).

Electric motors are key energy-intensive equipment in the industrial sector. Fig. 3.7 shows motor energy consumption for different applications in typical industries in Malaysia.

With factors determined from handheld instruments, the following equation is used to investigate the three-phase input power to the loaded motors. In addition, it is quantified the motor part load by the measuring input power

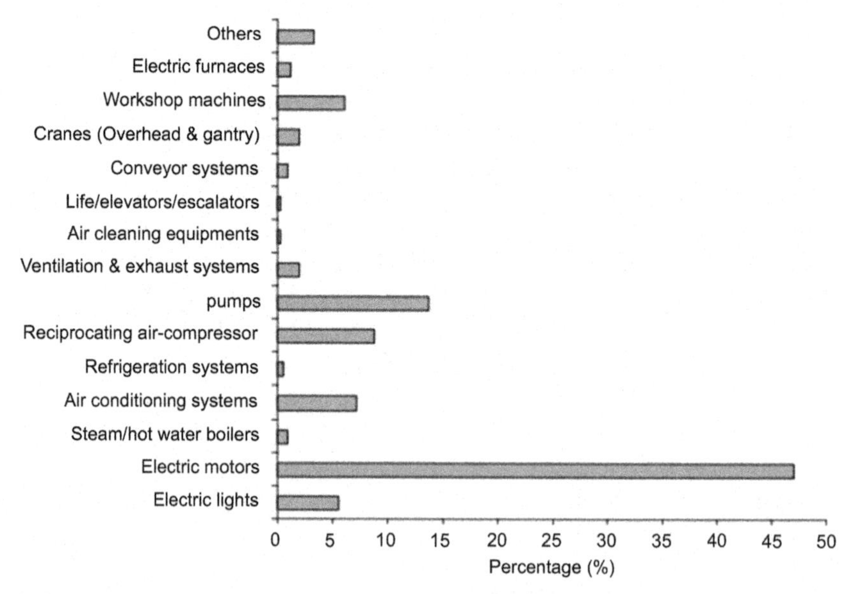

FIGURE 3.6 Energy audit on end use electricity consumption in Malaysian industries (R. Saidur et al., 2009).

TABLE 3.2 Motor energy uses by different countries.

Name of country	Motor energy usage (%)	Reference
Australia	30	(EEMDS, 2009)
Brazil	49	(Soares, 2007)
Canada	80	(Sterling, 1996)
China	60	(Yuejin, 2007)
EUEU	65–72	(Almeida, Fonseca, & Bertoldi, 2003; Tolvanen, 2008)
India	70	(Prakash, Baskar, Sivakumar, & Krishna, 2008)
Jordan	31	(Al-Ghandoor, Al-Hinti, Jaber, & Sawalha, 2008)
Korea	40	(KERI, 2007)
Malaysia	48	(Saidur et al., 2009)
Slovenia	52	(Al-Mansour, Merse, & Tomsic, 2003)
South Africa	60	(Khan, 2009)
Turkey	65	(Kaya et al., 2008)
UK	50	(Mecrow & Jack, 2008)
US	75	(Fatima Bouzidi, 2007; Lu, 2006)

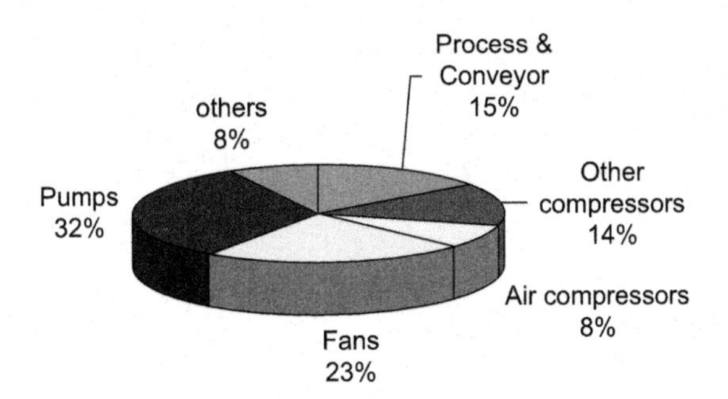

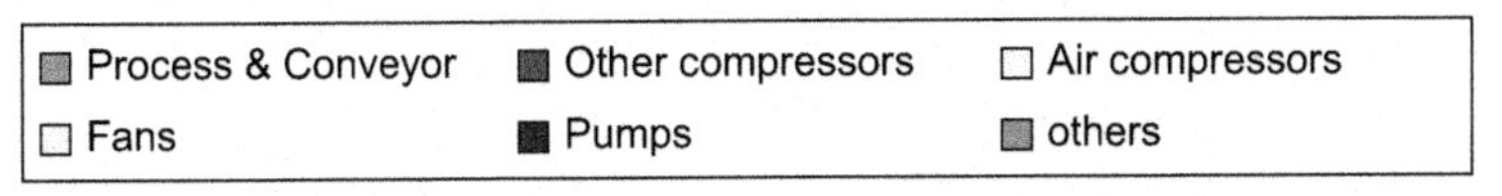

FIGURE 3.7 Motor energy consumption for different applications in plants (Cheng, 2003). (Hasanuzzaman, Rahim, Saidur, & Kazi, 2011).

compared under load to the power requires when the motors work at rated capacity (Fact sheet, 2014).

$$P_i = \frac{V \times I \times PF \times \sqrt{3}}{1000} \tag{3.36}$$

where

P_i = three-phase power, kW
PR = power factor as a decimal
I = RMS current, mean line to line of three phases, A
V = RMS voltage, mean of three phases, V

The power input at full rated load can be calculated as (Fact sheet, 2014)

$$P_{i\_F} = hp \times \frac{0.7457}{E_{motor\_F}} \tag{3.37}$$

where

P_{i_F} = power input at full rated load, kW
E_{motor_F} = efficiency at full rated load, %
0.746 = conversion from hp to kW
hp = motor rated horsepower

Moreover, the load factor, L, is power as a percentage of rated power, expressed as follows (Fact sheet, 2014):

$$L = \frac{P_i}{P_{i\_F}} \times 100 \tag{3.38}$$

where L is the load factor.

The yearly energy consumption attained by a motor can be estimated using the subsequent equations (Garcia, Szklo, Schaeffer, & McNeil, 2007; Habib, Hasanuzzaman, Hosenuzzaman, Salman, & Mehadi, 2016; Saidur, 2009; Saidur, Hasanuzzaman, Yogeswaran, Mohammed, & Shouquat, 2010):

$$AE = hp \times L \times 0.746 \times hr \times \left[\frac{1}{E_{motor}}\right] \times 100 \tag{3.39}$$

$$AE = P_m \times L \times hr \times \left[\frac{1}{E_{motor}}\right] \times 100 \tag{3.40}$$

where

AE is yearly energy consumption, kWh
hr is working hours, hr
E_{motor} is motor efficiency rating, %
P_m is motor rated power, kW
L is the load factor

3.5.3.2 Pump energy consumption

The variance head that the pumps have to connect consists of the vertical variance in height and the counterpressure. Moreover, a pump must connect the variance from the lower suction side pressure to the higher discharge side pressure, as shown in Fig. 3.8. As a result, the pressure variance, ΔP, through the pump is (Vogelesang, 2008)

$$\Delta P = P_d - P_s \tag{3.41}$$

where

P_d = higher discharge side pressure
P_s = lower suction side pressure

Suction side pressure can be determined as (Vogelesang, 2008)

$$p_s = p_{ts} + H_s \rho g \tag{3.42}$$

$$p_d = p_{td} + H_d \rho g \tag{3.43}$$

$$\Delta p = p_{td} + H_d \rho g - p_{ts} - H_s \rho g \tag{3.44}$$

$$\Delta p = (p_{td} - p_{ts}) + (H_d - H_s) \rho g \tag{3.45}$$

where

P_{td} = discharge tank pressure, N/m^2 (1 bar = 10^5 N/m^2)
H_s = fluid level height in suction tanks, m
P_{ts} = suction tank pressure, N/m^2
ρ = fluid density, kg/m^3

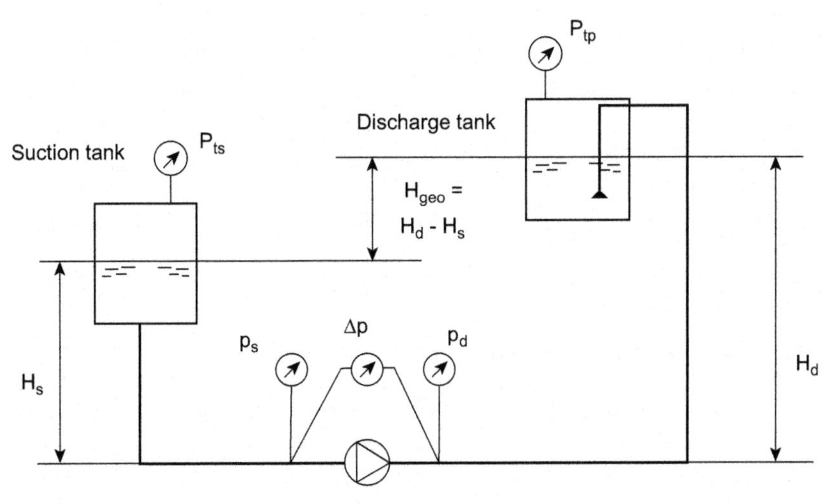

FIGURE 3.8 Vertical variance in height between discharge tanks and liquid (Vogelesang, 2008).

g = gravitational acceleration, 9.81 m/s^2
H_d = fluid level height in discharge tank, m

In the determination, H_d - H_s is the vertical variance in the height that must be connected between two fluids. This is called as the static (or geodetic) variance head and is commonly written as H_{geo}. The alteration pressure to be distributed through the pump is calculated by static differential head and by the change in the pressure of the two reservoirs, and is measured by applying the respective formula (Vogelesang, 2008):

$$\Delta p = (p_{td} - p_{ts}) + H_{geo}\rho g \tag{3.46}$$

The pump requires power to provide the differential/pressure head at a specific capacity that can be calculated as follows (Vogelesang, 2008):

$$P = (Q \times \Delta p)/36 \times \eta_{pump} \tag{3.47}$$

where

P = power to the shaft, kW
Δp = Bar
Q = fluid transmission capacity, m^3/h
η_{pump} = pump efficiency, %/100.

Then, the yearly energy consumption by an electric pump is measured as follows (Saidur, Hasanuzzaman, Yogeswaran, Mohammed, & Hossain, 2010; Saidur et al., 2012):

$$AEC = P \times hr \times 0.001 \tag{3.48}$$

where

AEC = yearly energy consumption, kWh
hr = annual usage hours
P = power, W

For instance, if
100 m^3/h = pump capacity,
1 bar = variable pressure, and
75% = efficiency, the output power required at the pump shaft will be (100×1)/(36×0.75) = 3.7 kW.

3.5.3.3 Compressed air

Compressed air is very essential for the industry, whether for running complicated works such as the process of pneumatic control or for simple tasks. Moreover, it is normally one of the very expensive values in an industrial facility. In midsize industry energy assessments, it is found that on average, compressed air makes up between 5% and 20% of the yearly electric cost of a plant.

The annual energy consumption attained by an air compressor is assessed as follows (Garcia et al., 2007; Saidur, 2009; Saidur et al., 2010):

$$AE = hp \times 0.746 \times hr \times \left[\frac{1}{E_{motor}}\right] \times 100 \quad (3.49)$$

where

AE = annual energy consumption, kWh
E = motor efficiency rating, %
hr = yearly working hours
hp = motor rated horsepower
0.746 = conversion factor for hp to kW

Then, the horsepower (hp) for a three-phase motor can be calculated as (Vogelesang, 2008)

$$hp = \frac{V \times I \times E_{motor} \times PF \times \sqrt{3}}{746} \quad (3.50)$$

where

V = RMS voltage, mean of three phases
PF = power factor as a decimal
I = RMS current, main line to line of three phases

The horsepower (hp) for a three-phase motor can be calculated as (Vogelesang, 2008):

$$hp = \frac{V \times I \times E_{motor} \times PF}{746} \quad (3.51)$$

where

I = current, A
PF = power factor as a decimal
V = voltage, V

3.5.3.4 Conveyor systems

Material transformation is done by conveyor belt. In addition, in the method of transferring material by conveyor belt, electrical energy is converted to movement energy, potential energy, heat energy, noise energy, etc. The model of energy conversion gives the connection between the conveyor parameters and energy. Conveyor parameter energy is the power included over a certainty of time. Energy consumption of a typical belt conveyor system, as shown in Fig. 3.9 can be divided into three categories:

- energy needs to move the vacant conveyor
- energy needs to lift material a definite height
- energy needs to run material horizontally over definite distance

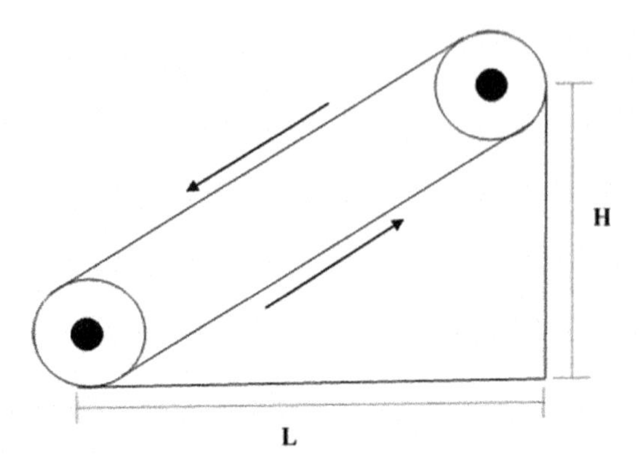

FIGURE 3.9 Schematic of a conveyor belt system

3.5.3.4.1 Energy to move the vacant conveyor

To move the vacant conveyor (without load condition), energy is needed to overcome resistance in the conveyor structures and to run the conveyor's different portions. The friction force (F_{ec}) or vacant conveyor is measured by the following formula (Halepoto & Khaskheli, 2016):

$$F_{ec} = gCQ(L + L_0) \tag{3.52}$$

where

F_{ec} = empty conveyor friction force, N
C = friction factor
$g = 9.81\ \text{m/s}^2$ = gravitational acceleration
Q = factor that represents the mass of the conveyor running portions for center to center distance, kg/m
L_o = terminal resistance independent of conveyor length or constant compensation length, m
L = horizontal projection of distance of decline or incline belts, m

The power (P_{ec}) to overcome this resistance force is measured by the following equations (Halepoto & Khaskheli, 2016):

$$P_{ec} = F_{ec} \times \frac{v}{1000} \tag{3.53}$$

$$P_{ec} = \frac{gCQ(L + L_0)v}{1000} \tag{3.54}$$

where

v = belt speed, ms^{-1}
P_{ec} = to move the vacant conveyor, kW

The energy consumption to run the empty conveyor is the power integral over time as follows (Halepoto & Khaskheli, 2016):

$$
\begin{aligned}
E_{ec} &= \int_0^t P_{ec} dt \\
&= \int_0^t \frac{gCQ(L+L_0)v}{1000} dt \\
&= \frac{gCQ(L+L_0)vt}{1000}
\end{aligned}
\tag{3.55}
$$

where

E_{ec} is the energy needed to run the empty conveyor (kWh)
t is period of belt operation (h)

3.5.3.4.2 Energy to run material horizontally

A load-conveyor belt practices an added resistance force owing to the load on the conveyor belt. This resistance force is measured by the following equations (Halepoto & Khaskheli, 2016):

$$F_h = gC(L+L_0) \times \left(\frac{T}{3.6 \times v}\right) \tag{3.56}$$

where

F_h is the load friction force to move material horizontally (N)
T is the transfer rate in tons per hour (t/h)

The power to move material parallel is obtained as follows (Halepoto & Khaskheli, 2016):

$$
\begin{aligned}
P_h &= F_h \times \frac{v}{1000} \\
&= gC(L+L_0) \times \left(\frac{T}{3.6 \times v}\right) \times \left(\frac{v}{1000}\right) \\
&= \frac{gC(L+L_0)T}{3600}
\end{aligned}
\tag{3.57}
$$

where

P_h is the power for conveying material horizontally (kW)

The energy is formed over the integral power result as follows:

$$E_h = \int_0^t P_h dt = \frac{gC(L+L_0)T}{3600} \times t \tag{3.58}$$

where

E_h is the energy to transfer material horizontally (kWh)
t is the period of belt operation (h)

3.5.3.4.3 Energy to raise or lower material

The perpendicular component of force lengthwise to raise or lower the load is measured by the following formula (Halepoto & Khaskheli, 2016):

$$F_l = \frac{gTH}{3.6 \times v} \tag{3.59}$$

where

F_l = component of force along the incline, N
T = transfer rate in tons per hour, t/h
$g = 9.8\ m/s^2$ = gravitational acceleration
H = net change in elevation, m
v = belt speed, ms^{-1}

The power can be calculated as:

$$P_l = = \frac{gTH}{3600} \tag{3.60}$$

where

P_l is the power to lift the load or the power generated by lowering the load (kW)

The applicable energy can be obtained as:

$$E_l = \frac{gTH}{3600} \times t \tag{3.61}$$

where

E_l is the energy to raise the load or the energy generated in lowering the load (kWh)

3.5.3.4.4 Total energy for the belt conveyor system

The overall energy consumption of a conveyor is the component summation of energy consumption and can be shown by the following equations (Halepoto & Khaskheli, 2016):

$$E_T = E_{ec} + E_h + E_l \tag{3.62}$$

where

E_t is total conveyor belt energy consumption (kWh)

3.5.3.5 Energy consumption of boiler

Boiler and other systems—for instance, ovens and furnaces—combust fuel with air for the object of releasing the heat energy of chemicals. Moreover, the object of the heat energy is to increase industrial product temperatures as a portion of manufacturing procedures. It may be used to produce high-pressure and high-temperature steam to power the output of a turbine or to basically heat a space to provide a feeling of comfort for its occupants (Hasanuzzaman et al., 2012).

$$\text{Energy\_consumption} = f(\text{load}) \times (1/\text{efficiency})dt$$

$$\text{Fuel energy} = \frac{\text{Boiler energy}}{\text{Combustion efficiency}}$$

$$\text{Boiler energy} = (\text{Steam energy flow} + \text{Feed water energy flow} + \text{Blowdown energy flow})$$

$$\text{Blowdown} = \text{Blowdown mass flow} + \text{specific enthalpy}$$

$$\text{Steam energy flow} = \text{Steam mass flow} + \text{specific enthalpy}$$

$$\text{Feed water} = \text{Feed water mass flow} + \text{specific enthalpy}$$

$$\text{Blowdown mass flow} = \text{Blowdown rate} \times \text{Feed water mass flow}$$

$$\text{Steam mass flow} = \text{Feed water massflow} - \text{Blowdown massflow}$$

$$\text{Steam mass flow} = \text{Feed water mass flow} - (\text{Blow down rate} \times \text{Feed water mass flow})$$

$$\text{Feed water mass flow} = \text{Steam mass flow}/(1 - \text{Blow down rate})$$

3.5.3.6 Boiler efficiency

Boiler efficiency is the efficiency of the combined result of the variance components of a boiler. Furthermore, the boiler has several subsystems whose efficiencies affect its total efficiency. Combined, the efficiencies considered in a boiler are:

- **thermal efficiency**
- **combustion efficiency**

The above efficiencies work with some other losses and play important roles and are important in considering the efficiency of the boiler and it is therefore essential that they be measured while computing boiler efficiency.

3.5.3.6.1 Combustion efficiency

Boiler combustion efficiency is the suggestion of burner capability to burn fuel. The burning efficiency depends on two different parameters such as excess O_2 levels in the exhaust and unburnt fuel quantity in the exhaust. The unburnt fuel quantity in the exhaust decreases with an increase in the excess air amount. In addition, this results in dropping the elevating the enthalpy losses with losing the unburnt fuel. Therefore, it is relatively important to conserve a balance between unburnt fuel losses and enthalpy losses. The combustion efficiency also differs with fuel existence burnt. The combustion efficiency is lower for the solid fuels than the gaseous and liquid fuels.

3.5.3.6.2 Thermal efficiency

Boiler heat exchanger effectiveness is specific boiler thermal efficiency, which is converted to heat energy from the fire side to the water side. In addition, thermal efficiency is seriously affected by soot formation/scale formation on the boiler tubes.

3.5.3.6.3 Indirect and direct boiler efficiency

The total efficiency of a boiler depends on several factors separately from the thermal and combustion efficiencies. These other factors consist of irradiation losses, off—on losses, blowdown losses, convection losses, etc. In actual exercise, two different procedures are frequently applied to find boiler efficiency, and are commonly known as indirect and direct processes of efficiency determination.

3.5.3.6.4 Direct efficiency

The efficiency formula has been used to calculate the boiler efficiency as follows (BEE, 2005; Patro, 2016):

$$\eta = (\text{output})/(\text{input}) \times 100 \tag{3.63}$$

Direct efficiency calculation (BEE, 2005; Patro, 2016):

$$\eta = \frac{Q \times (H - h)}{q \times GCV} \times 100 \tag{3.64}$$

where

H is steam enthalpy, kCal/kg
g is fuel quantity used per hour, kg/hr
Q is steam generated quantity, kg/hr
GCV is fuel gross calorific values
h is feed water enthalpy, kCal/kg

3.5.3.6.4.1 Indirect efficiency A boiler's indirect efficiency is determined by finding the individual losses enchanting in place in a boiler and then deducting those amounts from 100%. Moreover, this technique includes finding the magnitudes of entire calculable losses enchanting place in a boiler by individual measurements. The entire losses are added and deducted from

100% to realize the concluding efficiency that are calculated by using following equation (BEE, 2005; Patro, 2016).

$$\text{Indirect method boiler efficiency} = 100 - (\text{HL1} + \text{HL2} + \text{HL3} + \text{HL4} + \text{HL5} + \text{HL6} + \text{HL7}) \tag{3.65}$$

Where:

HL1- Loss owing to hydrogen in fuel, H_2
HL2- Loss owing to dry tube gas (sensible heat)
HL3 – Loss owing to moisture in fuel, H_2O
HL4 - Loss owing to carbon monoxide, CO
HL5 – Loss owing to surface irradiation, convection, and other unaccounted. The loss is unimportant and are problematic to determine
HL6- Loss owing to moisture in air, H_2O
In addition, the resulting losses are valid for solid-fired boilers:
HL7 - Bottom and fly ash un-burnt losses (Carbon)

1. Heat losses owing to flue gas

This is a key boiler loss and is calculated as follows (BEE, 2005; Patro, 2016):

$$\text{HL}_1 = \frac{m \times C_p(T_f - T_a)}{GCV_{fuel}} \times 100 \tag{3.66}$$

where

$L1$ = percentage of heat loss owing to dry flue gas
C_p = exact heat of flue gas, kCal/kg
T_a = ambient temperature, °C
T_f = temperature of flue gas, °C
GCV_{fuel} = fuel gross calorific value
m = mass of fuel dry flue gas, kg/kg
= fuel combustion products: $SO_2 + CO_2$ + nitrogen air provided definite mass + fuel nitrogen + flue gas of O_2

2. Heat losses owing to evaporation of water formed owing to hydrogen in fuel, %

H_2 combustion creates a heat loss, and the combustion product is water. The water is converted to steam, and this carries latent heat (Patro, 2016):

$$\text{HL}_2 = \frac{9 \times H_2 \times [584 + C_p(T_f - T_a)]}{GCV_{Fuel}} \times 100 \tag{3.67}$$

where

L_2 is the percentage of heat loss owing to the water evaporation

Tf is the temperature of flue gas, °C
C_p is the specific heat of superheated steam, kCal/kg °C
GCV_{fuel} is the fuel gross calorific value
Ta is the ambient temperature
584 is the latent heat constant for partial pressure of water steam

3. Heat loss owing to the moisture present in fuel

The fuel evacuates as a superheated vapor with moisture arriving at the boiler. Furthermore, this moisture loss is prepared as practical heat to pass the moisture to the boiling point and as superheated needs to get the steam to the exhaust gas temperature and the latent heat of the moisture evaporation. This moisture loss can be measured as follows (Patro, 2016):

$$\mathrm{HL}_3 = \frac{M \times [584 + C_p(T_f - T_a)]}{GCV_{Fuel}} \times 100 \tag{3.68}$$

where

L_3 is the percentage of heat loss owing to the moisture present in fuel
T_f is the temperature of flue gas, °C
C_p is the specific heat of the superheated steam, kCal/kg °C
M is the moisture mass in fuel in 1 kg basis
GCV_{fuel} is the fuel gross calorific value
Ta is the ambient temperature
584 is the latent heat consistent the partial pressure of water steam

4. Heat loss owing to the moisture present in the air

Vapor in the humidity form in the upcoming air, is super-heated as it passes over the boiler. Meanwhile, this heat passes up the chimney, it must be involved as a boiler loss (BEE, 2005; Patro, 2016):

$$\mathrm{HL}_4 = \frac{AAS \times f_{humidity} \times C_p \times (T_f - T_a)}{GCV_{Fuel}} \times 100 \tag{3.69}$$

where

L_4 is the percentage of heat loss owing to the air moisture present
T_f is the temperature of flue gas, °C
C_p is the specific heat of the steam superhated, kCal/kg°C
$F_{humidity}$ equal to kg of water/kg of dry air is the humidity factor
GCV_{fuel} is the fuel gross calorific value
Ta is the ambient temperature
AAS is the actual air mass supplied/kg of fuel

5. Heat loss owing to partial combustion

The products designed by partial combustion can be mixed with O_2 and burned again with further energy release. The different products include H_2, CO, and different hydrocarbons and are commonly initiated in the chimney gas of the boiler. CO is the only gas whose concentration can be measured suitably in a boiler plant investigation (Patro, 2016):

$$\mathrm{HL}_5 = \frac{\%CO \times C}{\%CO \times \%CO_2} \times \frac{5744}{GCV_{Fuel}} \times 100 \tag{3.70}$$

where

L_5 is the percentage of heat loss owing to incomplete conversion from carbon to CO
GCV_{fuel} is the fuel gross calorific value
C is the Carbon content of fuel, kg/kg
CO_2 is the CO_2 actual volume in flue gas, %
CO is the CO volume in flue gas leaving the economizer, %

6. Heat loss owing to irradiation and convection

Boiler heat losses by irradiation and convection from the boiler casting into the nearby the boiler house (Patro, 2016):

$$\mathrm{HL}_6 = 0.548 \times \left[(T_s/55.55)^4 - (T_a/55.55)^4\right] + 1.957 \times (T_s - T_a)^{1.25} \times \sqrt{[(196.85 \times V_m + 68.9)/68.9]} \tag{3.71}$$

where

HL_6 is the irradiation loss, W/m^2.
T_a is the ambient temperature, K
V_m is the wind velocity, m/s
T_s is the surface temperature, K

7. Heat loss owing to unburned carbon in bottom ash and fly ash

A slight amount of carbon will be absent in the ash, and this results in potential heat loss for the fuel. To calculate these heat losses, ash samples must be investigated for carbon content. The ash generated per unit quantity must also be identified. Heat loss owing to unburnt fly ash can be determined by the following formula (BEE, 2005; Patro, 2016):

$$\mathrm{HL}_7 = \frac{\text{Overall ash collected per kg of fuel burnt} \times \text{GCV fly ash}}{\text{fuel GCV}} \tag{3.72}$$

$$+\frac{\text{Overall ash collected per kg of fuel burnt} \times \text{GCV bottom ash}}{\text{fuel GCV}} \times 100$$

$$L_7 = \frac{\text{Overall ash collected per kg of fuel burnt} \times \text{GCV bottom ash}}{\text{fuel GCV}} \times 100$$

3.5.3.7 Energy intensity of industry

The energy efficiency in measured sectors could be conveyed in the energy-intensity form. The energy consumed or the energy intensity per unit of economic is commonly specified by value adding and is measured by (Saidur et al., 2007):

$$EI = \frac{TEC}{VA} \tag{3.73}$$

EI = energy Intensity
VA = products value addition from the industry annually
TEC = total energy (electricity and/or fuel) consumption in the industry annually

3.6 Transport sector

Transportation is a significant part of a national economy and its purposes as a foundation, service provider, and support. In addition, energy use in the transport sector contains the energy consumed in moving goods and people by air, rail, water, road, and pipeline. The Malaysian transportation sector consumed almost 43.3% of total energy usage of the country in 2013 (MESH, 2015).

The transportation sector comprises two major modes:

(i) passenger
(ii) freight

Passenger types include buses, trucks and light-duty cars, two- and three-wheel vehicles, passenger trains, and airplanes. Freight types are used in the movement of finished, intermediate, and raw goods to consumers and include light-duty, medium, and heavy trucks; international marine vessels and domestic marine vessels; pipelines; and rail.

3.6.1 Calculation based on transport energy rating

In this method, energy consumption is calculated according to the vehicle energy consumption in on road condition and this method is more reliable and appropriate. Different factors such as road condition, vehicle speed acceleration, and traffic are included in the energy consumption process.

3.6.1.1 Energy consumption by nonelectric road transportation sector

Specifically, the energy consumption of a vehicle consists of (1) accelerating resistance, (2) gradient resistance, (3) rolling resistance, and (4) air resistance. However, during traffic jams, a vehicle in pending position also consumes a trivial amount of energy that should be included.

So energy consumption by vehicles (De Cauwer, Van Mierlo, & Coosemans, 2015; Hyodo, Watanabe, & Wu, 2013) is calculated as

$$E_{Vehicle} = (m_{dw} + m_l)\cdot v\cdot a + \frac{1}{2}\rho\cdot C_D\cdot A.v^3 + (m_{dw} + m_l)\cdot g\cdot \nabla z + \mu_r \quad (m_{dw} + m_l)\cdot g\cdot v \tag{3.74}$$

where

m_{dw} = mass of vehicle (dead weight)
m_l = mass of load (passenger or freight)
v = velocity (m/s)
a = acceleration (m/s^2)
ρ = air density
A = vehicle projected area (m^2)
C_D = coefficient of air resistance
g = gravitational forces 9.8 (m/s^2)
Δz = difference of elevation in 1 s (m)
μ_r = rolling resistance coefficient

Energy consumption of vehicle without any passenger and/or freight load:

$$E'_{Vehicle} = m_{dw}\cdot v\cdot a + \frac{1}{2}\rho\cdot C_D\cdot A.v^3 + m_{dw}\cdot g\cdot \nabla z + \mu_r m_{dw}\cdot g\cdot v \tag{3.75}$$

Energy consumption by vehicle to cruise on level road:

$$E_{Vehicle} = (m_{dw} + m_l)\cdot v\cdot a + \frac{1}{2}\rho\cdot C_D\cdot A\cdot v^3 + \mu_r(m_{dw} + m_l)\cdot g\cdot v \tag{3.76}$$

Road load power—i.e., the minimum power required to overcome the impending condition of the vehicle (Heywood, 1988):

$$E_{RD} = \frac{1}{2}\rho\cdot C_D\cdot A\cdot v^3 + \mu_r(m_{dw} + m_l)\cdot g\cdot v \tag{3.78}$$

3.6.1.1.1 Instantaneous fuel consumption model

Instantaneous fuel consumption is shown by the fuel flow, Q = dC/dt (C is the consumed volume) expended by the vehicle's engine. Fuel flow, Q, is a function of acceleration, a, gear, G, and velocity, v. Therefore, fuel flow is $Q = (a, G, v)$. This can also be illustrated in terms of instantaneous energy consumption per distance, C_x by applying the chain rule as follows (Treiber, Kesting, & Thiemann, 2007):

$$C_x = \frac{dC}{dx} = \frac{1}{v}\frac{dC}{dt} = \frac{Q}{v} \tag{3.79}$$

The influencing aspects of instantaneous fuel consumption are as follows:

- vehicle acceleration (a), velocity (v)
- engine properties
- gear (G)
- state of engine and vehicle (e.g., engine cooler temperature) and external conditions (temperature, weather, road conditions, etc.)
- vehicle properties such as size, weight, transmission ratios, and coefficient (cd value)

3.6.1.1.2 Influence of engine properties on fuel consumption

Fuel consumption depends mainly on the revolution rate, f, and on effective motor pressure, p_e, of the car crankshaft. Additionally, using the volumetric caloric value, Δh_{vol}, of the fuel, which is 10.7 kWh/l or 38 MJ/L for diesel, and 10 kWh/l or 36 MJ/L for specific fuel consumption, C_{spec} (p_e, f), the fuel energy efficiency, γ (p_e, f), can be calculated as follows (Treiber, Kesting, & Thiemann, 2007):

$$\gamma(p_e, f) = \frac{1}{\Delta h_{vol} C_{spec}(p_e, f)} \tag{3.80}$$

However, the efficiency is exposed as the ratio between the enthalpy and the mechanical work of the fuel, where

$$\Delta H = \Delta h_{vol} C \tag{3.81}$$

The fuel rate can be presented in terms of the mechanical work and efficiency (Treiber, Kesting, & Thiemann, 2007),

$$Q = \frac{dc}{dt} \text{ and } P = \frac{dW}{dt}$$

$$Q(P, p_e, f) = \frac{P}{\Delta h_{vol} \gamma(p_e, f)} \tag{3.82}$$

Finally, the effective pressure, p_e, may be revealed in terms of mechanical power. For instance, for four-stroke engines, the relation is addressed by (Treiber, Kesting, & Thiemann, 2007)

$$p_e = \frac{2P}{V_{cyl} \times f} \tag{3.83}$$

where

V_{cyl} is the active volume of the entire cylinder engine.

3.6.1.1.3 Influence of car properties on fuel consumption

The relevant car properties determine how much mechanical power is required as a function of the vehicle's acceleration and velocity. The key effecting parameters are as follows (Treiber et al., 2007):

- electric consumers (such as air conditioning or lights) need a basis power, P_o
- aerodynamic drag, which leads to a force related to velocity squared
- power to overcome the rolling resistance and solid-state friction of tires
- power desired to overcome apathy force when accelerating, or gravitational drag when driving on an incline

3.6.1.2 Electric vehicle energy consumption calculation

Electric vehicles (EVs) indicate both battery-powered electric vehicles (BEVs) and plug-in hybrid electric vehicles (PHEVs). BEVs are totally battery dependent and provide whole petroleum displacement for definite vehicle sectors. PHEVs commonly have a reasonably sized energy storage system and internal combustion engine to confirm most miles are electrified while recalling the range capability of today's internal combustion engine vehicles. To recognize the EV's energy consumption characteristics, it is compulsory to recognize the relationships among velocity, power, roadway grade, and accelerating. Furthermore, the characteristic of EV energy consumption also depends on freeways and driving on urban streets. Referring to vehicle dynamics fundamental theory, the instantaneous power of EVs is measured by acceleration, vehicle speed, and roadway grade. Initially, the required tractive effort for an EV driving on definite conditions is measured by the key resistances as follows (Wu, Freese, Cabrera, & Kitch, 2015):

$$F = ma + R_a + R_r + R_g \tag{3.84}$$

where

F = tractive effort, lb or N,
a = acceleration, ft/s^2 or m/s^2,
m = vehicle mass, kg
R_g, R_{rl}, *and* R_a are grade, rolling, and aerodynamic resistances, respectively, ib or N
R_g, R_{rl}, and R_a can be determined as follows (Wu, Freese, Cabrera, & Kitch, 2015):

$$R_a = Kv^2 = \frac{\rho}{2} C_D A_f v^2 \tag{3.84}$$

$$R_{rl} = f_{rl} mg \tag{3.85}$$

$$R_g = mg \sin \theta \tag{3.86}$$

where

A_f = vehicle frontal area, m^2
C_D = drag coefficient
f_{rl} = rolling resistance constant
g = 32.2 ft/s^2 or 9.81 m/s^2 = gravity acceleration
k = aerodynamic resistance constant
v = velocity, m/s
θ^o = roadway grade
ρ = air density, kg/m^3

Thus

$$F = ma + Kv^2 + f_{rl}mg + mg \sin \theta \tag{3.87}$$

To produce tractive force, the required power for a vehicle traveling at velocity can be calculated as follows (Wu, Freese, Cabrera, & Kitch, 2015):

$$p = F \times v = \left(ma + Kv^2 + f_{rl}mg + mg \sin \theta\right) \times v \tag{3.88}$$

The p actually represents output power, which is provided by the input power (P). For EVs, P is produced by an electric motor. If it is assumed that the motor efficiency is η, then the following equation is the relationship between output power and input power:

$$p = \eta \times P \tag{3.89}$$

If we ignore the electricity used for climate control and other vehicle accessories, the majority of electrical power loss would be copper loss for iron loss (for an AC motor) or high current area (for a DC motor). Overall, the power losses can be defined as the product of the square of conductor resistance (r) and current (I). Therefore, motor efficiency η can be considered by:

$$\eta = \left(P - I^2 r\right)/P \tag{3.90}$$

where

r is conductor resistance, Ω
I is current, A

So the EV's instantaneous power can be assessed by the following equation:

$$P = I^2 r + F \times v \tag{3.91}$$

In contrast, the force (F) is produced by motor torque, which is abridged as the products of magnetic flux (Φ_d), armature constant (K_a), and current (I) (Wu, Freese, Cabrera, & Kitch, 2015):

$$F = \tau / R = (K_a \times \Phi_d \times I)/R \tag{3.92}$$

where

I = current, A
K_a = armature constant,
R = tire radius, m or ft
τ = torque, ib ft or Nm
Φ_d = Magnetic flux, weber

For AC and DC motors, Φ_d is different. For an AC motor, Φ_d is the RMS value of the direct axis air gap flux per pole; and for a DC motor, Φ_d is the direct axis air gap flux per pole. Finally, combining all equations, EV instantaneous power is assessed by (Wu, Freese, Cabrera, & Kitch, 2015):

$$P = \frac{r \times R^2}{(K_a \times \Phi_d)^2} \times \left(ma + Kv^2 + f_{rl}mg + mg \sin\theta\right)^2 + \left(Kv^2 + f_{rl}mg + mg \sin\theta\right) \times v + mav \tag{3.93}$$

This equation is shortened as follows (Wu, Freese, Cabrera, & Kitch, 2015):

$$P = P_m + P_t + P_g \tag{3.94}$$

where

$P_m = \frac{r \times R^2}{(K_a \times \Phi_d)^2} \times \left(ma + Kv^2 + f_{rl}mg + mg \sin\theta\right)^2$ = motor power losses;

$P_t = (Kv^2 + f_{rl}mg + mgsin\theta) \times v$ = power losses caused by travel resistance; and

$P_g = mav$ = acceleration possible gained energy.

3.6.2 Calculation based on transport activity

In lack of energy data, it is probable to calculate the average energy consumption from the activity of the transport that means from the travel distance, number of passenger or amount of goods carried etc. Energy consumption varied according to transportation types such as motorcycles, trucks, cars, motorcycles, light vehicles, air transport, buses, water transport, and rails etc. Overall, transportation can be classified as:

(i) electric
(ii) nonelectric.

3.6.2.1 Energy consumption by nonelectric vehicle

Generally, all forms of energy, either electric or nonelectric, are converted into tonnes of oil equivalent (toe). So for nonelectric vehicles, first fuel consumption is calculated and then the consumed fuel needed to convert into toe. The general formula of fuel consumption of a nonelectric vehicle is as follows:

$$F_C = N_T \times L_D \times F_{Fuel} \tag{3.95}$$

where

F_C = fuel consumption by transportation sector
N_T = number or volume of transport by mode of transport
E_{Fuel} = average fuel economy in liters per km or (tonne-km) or (passenger-km) by transport mode
L_D = average transport distance by mode of transport

To get the actual energy consumption of a particular vehicle, the consumed fuel is needed to convert in to energy. The energy consumption formula is as:

$$E_C = F_C \times C_f \tag{3.96}$$

where

E_C = energy consumption
F_C = fuel consumption
C_f = fuel-to-energy conversion factor

The fuel-to-energy conversion factors are adopted from the *Units and Conversion Fact Sheet of MIT Energy Club.* Energy content in different fuels are as follows (MITLE, 2019):

Biodiesel	124.80 MJ/gal	38.00 MJ/L	37.50 MJ/kg
CNG @ 20 MPa	50.00 MJ/kg	9.30 MJ/L	248.60 mBtu/ft^3
Crude oil	145.70 MJ/gal	38.50 MJ/L	43.80 MJ/kg
Diesel	135.50 MJ/gal	35.80 MJ/L	42.80 MJ/kg
Ethanol	80.20 MJ/gal	21.20 MJ/L	26.90 MJ/kg
Gasoline	121.30 MJ/gal	32.10 MJ/L	43.10 MJ/kg
H_2 @ 35 MPa (HHV)	120.00 MJ/kg	2.70 Mj/L	72.50 mBtu/ft^3
LPG @ 1.5 MPa	88.10 MJ/gal	23.30 MJ/L	625.50 mBtu/ft^3
Methanol	60.40 MJ/gal	15.90 MJ/L	20.10 MJ/kg
Nat gas @ STP	53.20 MJ/kg	38.20 MJ/m^3	1027 Btu/ft^3

Note. 1 Btu = 1055 J. 1 kWh = 3.6 MJ = 3412 Btu. 1 hp = 746 W. 1 TW ≈ 30 Quad/yr ≈ 32 EJ/yr. 1 quad = 10^{15} Btu ≈ 1.05 EJ ≈ 25 Mtoe. 1 L diesel = 35.8 MJ = 35.8 × 10^{-6} TJ = 35.8 × 10^{-6} ÷ 41.868 ktoe = 8.55 × 10^{-7} ktoe. 1 L petrol = 32.1 MJ = 32.1 × 10^{-6} TJ = 32.1 × 10^{-6} ÷ 41.868 ktoe = 7.67 × 10^{-7} ktoe. 1 L NGV = 9.3 MJ = 9.3 × 10^{-6} TJ = 9.3 × 10^{-6} ÷ 41.868 ktoe = 2.22 × 10^{-7} ktoe. 1 toe = 41.868 GJ; i.e., 1 ktoe = 41.868 × 10^3 GJ = 41.868 TJ.

Fuel economy depends on the transport type, size, engine capacity, and load. Fuel economy is measured by the specific fuel consumption, expressed as: unit consumption per 100 km for motorcycle or private car; per pass-km for passenger taxi, bus, or rail, per passenger per ton-km for goods vehicles and for air transport, etc. The average fuel economies of different transportation modes according to their engine capacities are presented in the following (Tables 3.3-3.6):

3.6.2.2 Electric vehicle energy consumption

On the other hand, nowadays EVs also contribute to the transportation sector, and EV demand is growing due to green environmental impacts. The energy consumption calculation for EVs can be done as follows:

$$E_{EV} = N_{EV} \times L_D \times E_{Economy} \tag{3.97}$$

where

TABLE 3.3 Average fuel economy of a motorcycle.

Engine capacity	Fuel type	Mileage (km/L)
≤ 100 cc	Petrol	45–60
101–150 cc	Petrol	40–45
151–250 cc	Petrol	30–40
> 250 cc	Petrol	20–30

TABLE 3.4 Average fuel economy of a car.

Engine capacity	Fuel type	Mileage (km/L)
≤ 1000 cc	Petrol	16–20
1001–1500 cc	Petrol	12–16
1501–2000 cc	Petrol	8–12
> 2000 cc	Petrol/diesel	< 8

TABLE 3.5 Average fuel economy of a bus.

Engine type	Fuel type	Mileage (km/L)
≤ 1500 cc	Diesel	6–8
1501–4000 cc	Diesel	4–6
> 4000 cc	Diesel	< 4

TABLE 3.6 Average fuel economy of goods vehicles.

Engine capacity	Fuel	Mileage (km/L)
$\leq$ 2000 cc	Diesel/petrol	8–10
2001–4000 cc	Diesel	5–8
4001–10000 cc	Diesel	3–5
> 10000 cc	Diesel	< 3

E_{EV} = energy consumption of electric vehicle
N_{EV} = number of electric vehicles
L_D = distance traveled by a single electric vehicle
$E_{Economy}$ = energy consumption by a single electric vehicle per unit of travel distance

References

Ahmed, S. S., Iqbal, A., Sarwar, R., & Salam, M. S. (2014). Modeling the energy consumption of a lift. *Energy and Buildings, 71*(Suppl. C), 61–67. https://doi.org/10.1016/j.enbuild.2013.12.005.

Al-Ghandoor, A., Al-Hinti, I., Jaber, J. O., & Sawalha, S. A. (2008). Electricity consumption and associated GHG emissions of the Jordanian industrial sector: Empirical analysis and future projection. *Energy Policy, 36*(1), 258–267.

Al-Mansour, F., Merse, S., & Tomsic, M. (2003). Comparison of energy efficiency strategies in the industrial sector of Slovenia. *Energy, 28*(5), 421–440.

Almeida, A. T., Fonseca, P., & Bertoldi, P. (2003). Energy-efficient motor systems in the industrial and in the services sectors in the European Union: Characterisation, potentials, barriers and policies. *Energy, 28*(7), 673–690.

Barney, G., & Lorente, A. (2013). Simplied energy calculations for lifts based on ISO/DIS 25745-2. In *Presented at the 3rd symposium of lift and escalator technologies, Northampton, UK, sept. 26-27, 2013.*

BCIS. (2013). *Method to Identify Building Energy Index (BEI), NET BEI, GFA, NFA, ACA in several projects in Malaysia since 2000 (including KeTTHA and agencies), Malaysian Green Technology Corporation (GreenTech Malaysia) and Sustainable Energy Development Authority (SEDA) Malaysia, May 2013, seminar by building consumption input system.* https://efit.seda.gov.my/?omaneg...id=2738(20/07/2019).

BEE. (2005). *Energy performance assessment of boilers.* Retrieved from https://beeindia.gov.in/sites/default/files/4Ch1.pdf.

Bouzidi, F. (2007). *Energy savings on single-phase induction motors under light load conditions.* USA: MS thesis, University of Nevada Las Vegas.

Brady, L., & Abdellatif, M. (2017). Assessment of energy consumption in existing buildings. *Energy and Buildings, 149*(Suppl. C), 142–150. https://doi.org/10.1016/j.enbuild.2017.05.051.

Chan, S. A. (2009). *Green building index – MS1525.* Retrieved from http://new.greenbuildingindex.org/Files/Resources/20090214%20-%20GBI%20MS1525-2007%20Seminar/20090214%20-%20GBI%20MS1525-2007%20Seminar%20(CSA)%20Notes.pdf.

Cheng, C. S. (2003). *High efficiency motors for industrial and commercial sectors in Malaysia.* http://www.ptm.org.my/mieeip/pdf.

Chua, S. C., & Oh, T. H. (2010). Review on Malaysia's national energy developments: Key policies, agencies, programmes and international involvements. *Renewable and Sustainable Energy Reviews, 14*(9), 2916–2925. https://doi.org/10.1016/j.rser.2010.07.031.

Chua, S. C., & Oh, T. H. (2011). Green progress and prospect in Malaysia. *Renewable and Sustainable Energy Reviews, 15*(6), 2850–2861. https://doi.org/10.1016/j.rser.2011.03.008.

De Cauwer, C., Van Mierlo, J., & Coosemans, T. (2015). Energy consumption prediction for electric vehicles based on real-world data. *Energies, 8*(8), 8573.

EEMDS. (2009). http://www.controleng.com/article/269285-Energy_Efficient_Motors_Deliver_Savings.php.

EIA. (2016). *International energy outlook 2016.* Retrieved from https://www.eia.gov/outlooks/ieo/pdf/0484(2016).pdf.

Fact sheet. (2014). *DETERMINING ELECTRIC MOTOR LOAD AND EFFICIENCY.* U.S. Department of Energy. Retrieved from https://www.energy.gov/sites/prod/files/2014/04/f15/10097517.pdf.

Garcia, A. G. P., Szklo, A. S., Schaeffer, R., & McNeil, M. A. (2007). Energy-efficiency standards for electric motors in Brazilian industry. *Energy Policy, 35*(6), 3424–3439.

Habib, M. A., Hasanuzzaman, M., Hosenuzzaman, M., Salman, A., & Mehadi, M. R. (2016). Energy consumption, energy saving and emission reduction of a garment industrial building in Bangladesh. *Energy, 112*, 91–100. https://doi.org/10.1016/j.energy.2016.06.062.

Halepoto, I. A., & Khaskheli, S. (2016). Modeling of an Integrated Energy Efficient Conveyor System Model using Belt Loading Dynamics. *Indian Journal of Science and Technology, 9*(47). https://doi.org/10.17485/ijst/2016/v9i47/108658.

Heywood, J. B. (1988). *Internal combustion engine fundamentals.* McGraw-Hill.

Hasanuzzaman, M., Rahim, N. A., Saidur, R., & Kazi, S. N. (2011). Energy savings and emissions reductions for rewinding and replacement of industrial motor. *Energy, 36*(1), 233–240. https://doi.org/10.1016/j.energy.2010.10.046.

Hasanuzzaman, M., Rahim, N. A., Hosenuzzaman, M., Saidur, R., Mahbubul, I. M., & Rashid, M. M. (2012). Energy savings in the combustion based process heating in industrial sector. *Renewable and Sustainable Energy Reviews, 16*(7), 4527–4536. https://doi.org/10.1016/j.rser.2012.05.027.

Hyodo, T., Watanabe, D., & Wu, M. (2013). Estimation of energy consumption equation for electric vehicle and its implementation. In *Paper presented at the 13th World conference on transportation research Rio de Janeiro Brazil.*

IEA. (2017). *Key world energy statistics 2017.* International Energy Agency (IEA). https://www.iea.org/publications/freepublications/publication/KeyWorld2017.pdf.

IEO. (2016). *Industrial sector energy consumption.* U.S. Energy Information Administration | International Energy Outlook, 2016 https://www.eia.gov/outlooks/ieo/pdf/industrial.pdf.

ISO. (2013). *ISO Standard 12655: Energy performance of buildings – presentation of real energy use of buildings*, 2013.

Kaya, D., Yagmur, E. A., Yigit, K. S., Kilic, F. C., Eren, A. S., & Celik, C. (2008). Energy efficiency in pumps. *Energy Conversion and Management, 49*(6), 1662–1673.

KERI. (2007). *Korea electrotechnology research institute, minimum energy performance standards for three phase induction motors in Korea energy efficiency in motor driven systems EEMODS 2007 held in Beijing, China, from June 10-13.*

Khan, A. (2009). *Electric motor efficiency, international conference on industrial and commercial use of energy (ICUE, 09).* June 10-11, South Africa.

Khan, A. (2019). *Khan Academy, Price elasticity of demand and price elasticity of supply.* https://www.khanacademy.org/economics-finance-domain/microeconomics/elasticity-tutorial/price-elasticity-tutorial/a/price-elasticity-of-demand-and-price-elasticity-of-supply-cnx (20/07/2019).

KENTON, W. (2019). *Demand Elasticity.* https://www.investopedia.com/terms/d/demand-elasticity.asp(20/07/2019).

Li, J. Z. (2007). *Study on the life cycle consumption of energy and resource of air conditioning in urban residential buildings in China.* China: Tsinghua University (Ph. D.).

Li, C., Tang, S., Cao, Y., Xu, Y., Li, Y., Li, J., & Zhang, R. (2013). A New Stepwise Power Tariff Model and Its Application for Residential Consumers in Regulated Electricity Markets. *IEEE Transactions on Power Systems, 28*(1), 300–308. https://doi.org/10.1109/TPWRS.2012.2201264.

Lu, B. (2006). *Energy usage evaluation and condition monitoring for electric machines using wireless sensor networks, PhD thesis.* USA: Georgia Institute of Technology.

Marx, D. J. L., & Calmeyer, J. E. (2004). A case study of an integrated conveyor belt model for the mining industry, 15-17 Sept. 2004. In *Paper presented at the 2004 IEEE Africon. 7th Africon conference in Africa (IEEE cat. No.04CH37590).*

McKinnon, A. C., & Piecyk, M. (2010). *Measuring and managing CO2 emissions in European chemical transport* (Retrieved from).

MEAC. (2014). *Methodology for calculating the energy performance of buildings. Riigi Teataja.* Retrieved from https://www.riigiteataja.ee/en/eli/520102014002/.

Mecrow, B. C., & Jack, A. G. (2008). Efficiency trends in electric machines and drives. *Energy Policy, 36*(12), 4336–4341.

MESH. (2015). *Malaysia energy statistics, handbook 2015, suruhanjaya tenaga (energy commission).*

MITEL. (2019). *Units & Conversion Fact Sheet of MIT Energy Club.* tps://cngcenter.com/wp-content/uploads/2013/09/UnitsAndConversions.pdf(20/07/2019).

Ouf, M. M., & Issa, M. H. (2017). Energy consumption analysis of school buildings in Manitoba, Canada. *International Journal of Sustainable Built Environment, 6*(2), 359–371. https://doi.org/10.1016/j.ijsbe.2017.05.003.

Prakash, V., Baskar, S., Sivakumar, S., & Krishna, K. S. (2008). A novel efficiency improvement measure in three-phase induction motors, its conservation potential and economic analysis. *Energy for Sustainable Development, 12*(2), 78–87.

Saidur, R. (2009). Energy consumption, energy savings, and emission analysis in Malaysian office buildings. *Energy Policy, 37*(10), 4104–4113.

Saidur, R., Hasanuzzaman, M., Sattar, M. A., Masjuki, H. H., Irfan Anjum, M., & Mohiuddin, A. K. M. (2007). An analysis of energy use, energy intensity and emissions at the industrial sector of Malaysia. *International Journal of Mechanical and Materials Engineering, 2*(1), 84–92.

Saidur, R., Hasanuzzaman, M., Yogeswaran, S., Mohammed, H. A., & Hossain, M. S. (2010). An end-use energy analysis in a Malaysian public hospital. *Energy, 35*(12), 4780–4785.

Saidur, R., Rahim, N. A., Masjuki, H. H., Mekhilef, S., Ping, H. W., & Jamaluddin, M. F. (2009). End-use energy analysis in the Malaysian industrial sector. *Energy, 34*(2), 153–158. https://doi.org/10.1016/j.energy.2008.11.004.

Saidur, R., Sambandam, M., Hasanuzzaman, M., Devaraj, D., Rajakarunakaran, S., & Islam, M. (2012). An energy flow analysis in a paper-based industry. *Clean Technologies and Environmental Policy, 14*(5), 905–916. https://doi.org/10.1007/s10098-012-0462-9.

SEA. (2018). *Energy intensive industry.* Swedish Energy Agency. http://www.energimyndigheten.se/en/innovations-r–d/energyintensive-industry/(05/02/2018.

Sharif, L. A.-. (1996). Lift & escalator energy consumption. In *Proceedings of the CIBSE/ASHRAE joint national conference, Harrogate, UK, 29 Sept. - 1 Oct. 1996* (Vol. 1, pp. 231–239).

Soares, G. A. (2007). Brazilian industrial energy efficiency program: Focus on motor driven systems. In *Energy efficiency in motor driven systems EEMODS 2007 held in Beijing, China, from June 10-13.* http://www.controleng.com/article/269285-Energy_Efficient_Motors_Deliver_Savings.php, 29/08/09.

Sterling, C. A. (1996). *Prevalence of components necessary for electrical demand side management savings persistence in the Albertan Industrial market sector.* MS thesis. Canada: University of Alberta.

Tolvanen, J. (2008). *LCC approach for big motor-driven systems savings* (pp. 24–27). November: World Pumps.

Treiber, M., Kesting, A., & Thiemann, C. (2007). *How much does traffic congestion increase fuel consumption and emissions? Applying a fuel consumption model to the NGSIM trajectory data.* http://www.mtreiber.de/publications/fuelModel.pdf (16/11/2017).

Tukia, T. (2014). *Determining and modeling the energy consumption of elevators, Thesis submitted for examination for the degree of Master of Science in Technology.* School of Electrical Engineering, Aalto University.

Uimonen, S., Tukia, T., Siikonen, M.-L., Lange, D., Donghi, C., Cai, X. L., … Lehtonen, M. (2016). Energy consumption of escalators in low traffic environment. *Energy and Buildings, 125*(Suppl. C), 287–297. https://doi.org/10.1016/j.enbuild.2016.05.018.

Vogelesang, H. (2008). An introduction to energy consumption in pumps. *World Pumps, 2008*(496), 28–31. https://doi.org/10.1016/S0262-1762(07)70434-0.

Wei, Y., Zhang, X., Shi, Y., Xia, L., Pan, S., Wu, J., … Zhao, X. (2018). A review of data-driven approaches for prediction and classification of building energy consumption. *Renewable and Sustainable Energy Reviews, 82*, 1027–1047. https://doi.org/10.1016/j.rser.2017.09.108.

Wu, X., Freese, D., Cabrera, A., & Kitch, W. A. (2015). Electric vehicles' energy consumption measurement and estimation. *Transportation Research Part D: Transport and Environment, 34*(Suppl. C), 52–67. https://doi.org/10.1016/j.trd.2014.10.007.

Yang, Q., Liu, M., Huang, C., Min, Y., & Zhong, Y. (2015). A model for residential building energy consumption characteristics and energy demand: A case in Chongqing. *Procedia Engineering, 121*(Suppl. C), 1772–1779. https://doi.org/10.1016/j.proeng.2015.09.154.

Yuejin, Z. (2007). General situation of energy conservation standards for China's motor system. In *Energy efficiency in motor driven systems EEMODS 2007 held in Beijing, China, from June 10-13.*

Zhao, D., McCoy, A. P., Du, J., Agee, P., & Lu, Y. (2017). Interaction effects of building technology and resident behavior on energy consumption in residential buildings. *Energy and Buildings, 134*, 223–233. https://doi.org/10.1016/j.enbuild.2016.10.049.

Chapter 4

Energy supply

M. Hasanuzzaman, Laveet Kumar
Higher Institution Centre of Excellence (HICoE), UM Power Energy Dedicated Advanced Centre (UMPEDAC), Level 4, Wisma R&D, University of Malaya, Jalan Pantai Baharu, Kuala Lumpur, Malaysia

4.1 Introduction

With growth in global population and rapid economic development, the energy issue has increasingly become a grave human concern. Due to the limited resource of traditional energy and its negative impact on the environment, humans must explore other renewable, environmentally friendly, low-carbon energy types. Renewable energy sources such as wind, solar, and biomass have become important in human development due to their sustainable use and environmentally friendly characteristics. The development and utilization of renewable energy (RE) are a trend in modern society. The vigorous growth of RE has become an essential part of many countries' energy strategies. RE utilization can be divided into three categories: power generation, transportation, and heating and cooling. By the end of 2017, 128 countries had developed appropriate regulatory incentives or mandatory policies for RE power generation but rarely countries have policies related to industries such as transportation, heating, and cooling. In RE power generation, hydropower is dominant, followed by wind power and photovoltaic. Others such as biomass power, geothermal power, solar thermal power, and ocean energy generation only occupy a secondary position.

According to the REN21 (Renewable Energy Policy Network for the 21st Century) report 2017, global RE investment increased by 4.4 times from 2004 to 2016, with China being the country with the most significant investment in renewable power generation and renewable fuels. As of 2016, global RE power generation accounted for 24.5% of total power generation, of which hydropower generation was the highest, accounting for 16.6%, followed by wind farms accounting for 4.0%, biomass power generation accounting for 2.0%, and photovoltaic power generation accounting for 1.5%. The scope of promotion and application of RE is expanding rapidly.

Energy for Sustainable Development. https://doi.org/10.1016/B978-0-12-814645-3.00004-3

With increased use of RE, related research in the energy supply chain (ESC) has also been valued by society, such as interest in the workings of RESC process flow. What are the indicators of sustainable development? Degrees of fairness are taken into account to ensure affordability in the deployment of RE of developed and developing countries, which have different energy consumption abilities. Development and transportation of RE may require new facilities, judgment of economic viability, and the need to measure the process from RE delivery to ensure that the end user is environmentally friendly—according to the World Nuclear Energy Association, in the horizontal comparison of power generation, wind, solar, biomass, and other RE generation life cycle GHG emissions are significantly lower than they are for thermal power generation, and the emission reduction effect is very significant. There is no doubt that it is an environmentally friendly clean energy source. However, if we further analyze the life cycle from the perspectives of planning, design, production, operation, end use, maintenance and recycling, the cleanliness of RE power generation will also have hidden emissions that cannot be ignored. In the case of wind and power, it has zero direct carbon emissions during operation, but some products or services required for wind farm operation need to be obtained from other departments, such as blades, generators, gearboxes, and control systems, which must be purchased from the relevant fan equipment supplier, and they may cause dioxide emissions (Xu, Pang, Zhang, Poganietz, & Marathe, 2018). And because most REs are unstable natural resources, how to examine their risk during the whole life cycle is an important question. Besides, energy dependence and energy supply diversity are essential research topics because they have a significant impact on national energy strategies, based on the above considerations and the use of scientific indicators to conduct energy forecasting.

4.2 Energy supply

Supply is the quantity of something that producers have available for sale. Energy supply is the quantity of energy that suppliers/resources have available to provide to end users. Fig. 4.1 shows the energy supply and demand structure of the residential sector (Han & Kim, 2017). The wide range of energy sources and carriers that provide energy services must

- offer long-term security of supply,
- be affordable,
- have minimal impact on the environment,
- provide security of energy supply issues,
- add perceived future benefits, and
- lower carbon-emitting technologies.

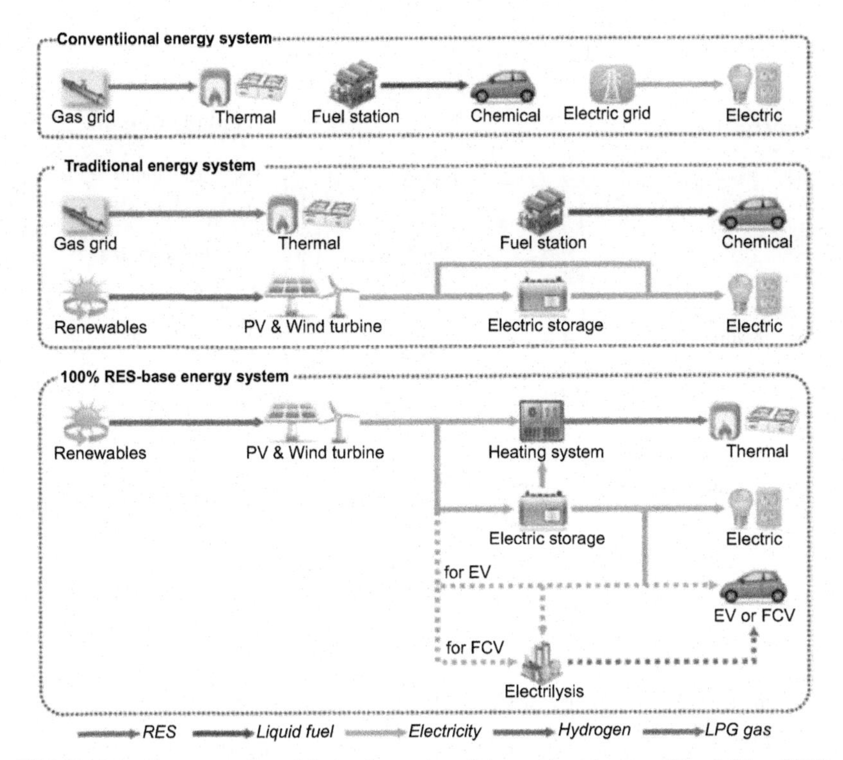

FIGURE 4.1 Energy supply and demand structure of the residential sector (Han & Kim, 2017).

4.3 Renewable energy supply chain

Energy supply is the delivery of fuel or transformed fuel to the point of consumption. Fig. 4.2 shows the energy input and output scheme (Feng, Mears, Beaufort, & Schulte, 2016). The process involves extraction, transmission, generation, distribution, and finally storage of fuels, and is often referred to as energy flow. The ESC consists of three major components; (1) energy commodity, (2) network services, and (3) retail services (Fig. 4.2).

Energy commodity is electricity generation at the power plant or the natural gas production, for example. The transmission or shifting of energy in high amounts/volumes from their origin toward the end user falls under network services. Retail services such as billing and price risk management for the end user are the final component of ESC. These components may be managed by different entities or businesses.

While RE resources are enormous, their supply is unpredictable, as most RE forms are nature/weather dependent. Redistribution and utilization form the major components of the RE supply chain.

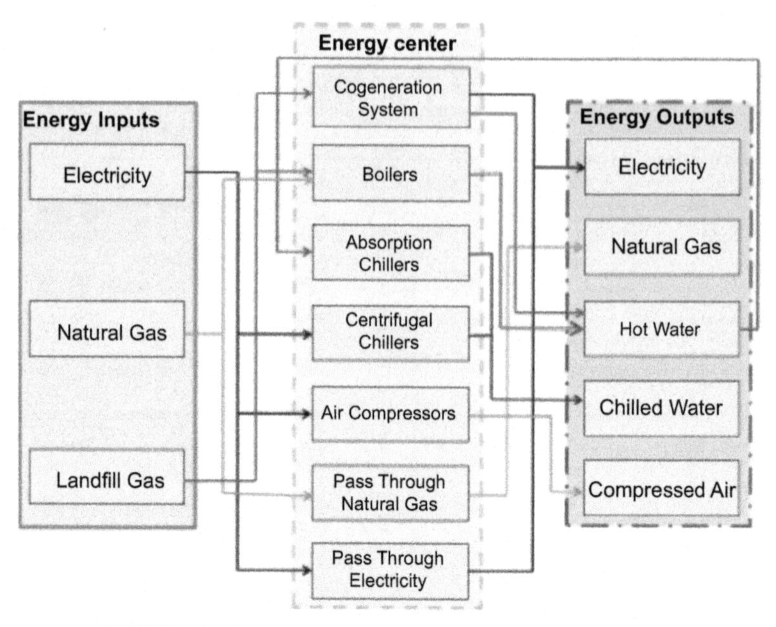

FIGURE 4.2 Energy input and output scheme (Feng et al., 2016).

4.3.1 Renewable energy supply chain process flow

Renewable energy supply chain (RESC) elements includes physical, information (technical and general), and financial flows. As an example, to demonstrate the relationships, electricity is portrayed in the supply chain process flow. RESC flow is shown in Fig. 4.3, as presented by the United Nations Development Program. It must be noted that technology is a critical factor for improving the efficiency and originating the distribution network. Replacement of conventional energy sources with RE is possible only by the commercialization of RE.

4.3.2 Renewable energy supply chain limitations and issues

Renewable energy sources are limited by essential characteristics just like other conventional energy sources. RE resources have three critical parameters—i.e., intermittency, variability, and maneuverability. These critical parameters require effective management and control. The concerns and issues of RESC are shown in Fig. 4.4 (Evans, Strezov, & Evans, 2009; Shen, Lin, Li, & Yuan, 2010).

4.4 Energy supply indicators for sustainable development

Utilization of RE contributes to the national energy supply security and socioeconomic development of a country. Therefore, different types of energy

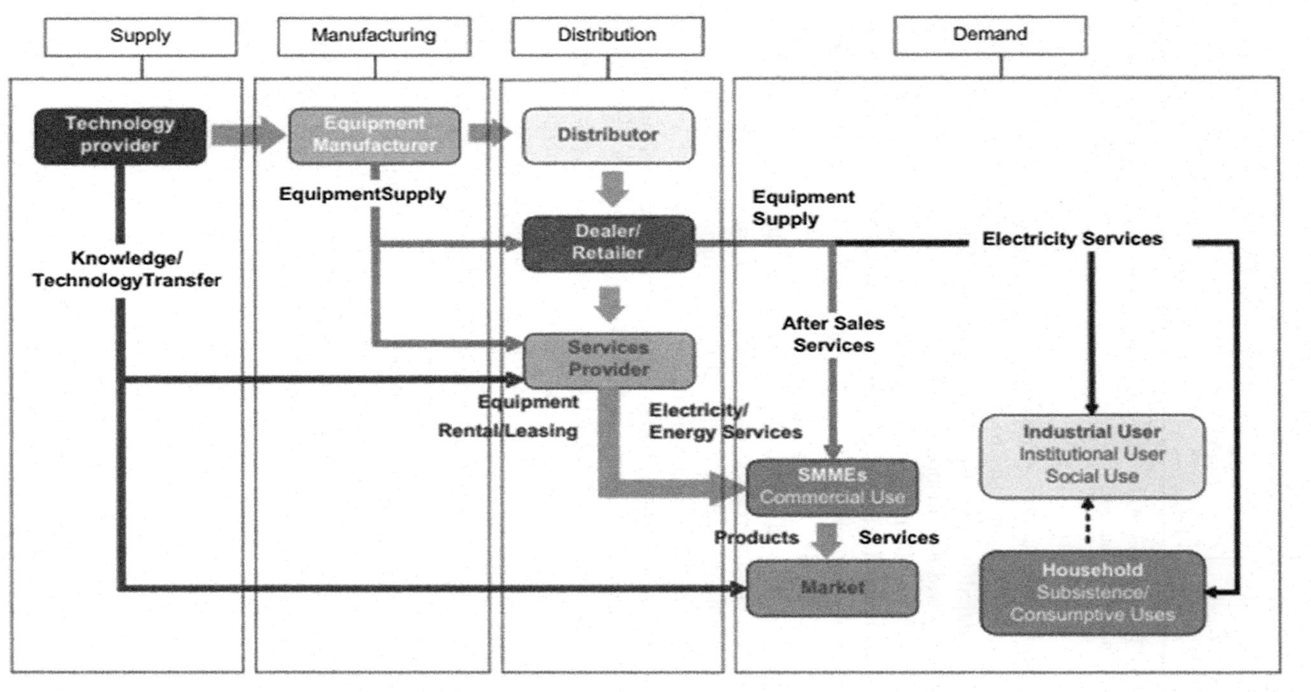

FIGURE 4.3 Renewable energy supply chain process (UNDP, 2010).

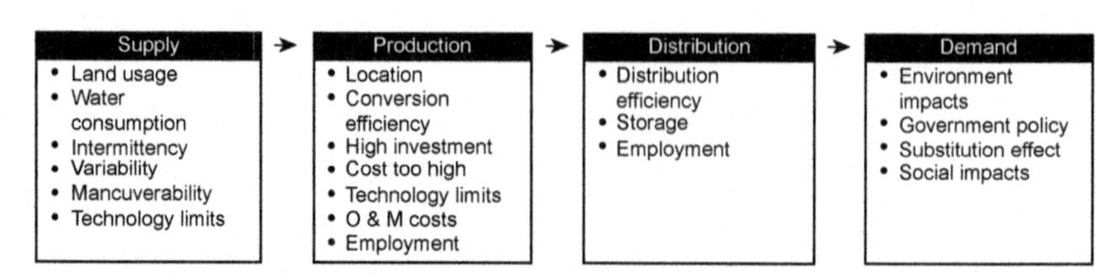

FIGURE 4.4 The concerns of the renewable energy supply chain. (Evans, Strezov, & Evans, 2009; Shen, Lin, Li, & Yuan, 2010).

supply policies and action plans have been implemented to facilitate the growth of the RE industry and to reduce the severe dependence on electricity generation created by the use of fossil fuels. The International Renewable Energy Agency was established to enhance deployment of off-grid RE through strategic policies planning, appropriate financial schemes assistance or offering, and technologies advancement and knowledge transfer as well as innovative business models and investment, which lead to a sustainable energy future (IRENA, 2018).

The progress toward sustainable development in energy by deploying RE should be tracked, and regular monitoring should be done on the impacts of implemented policies and strategies. It is also crucial to understand and give concerns about the consequences of the entire RESC life cycle, from the manufacturing or construction stage to electricity generation and distribution, and lastly to waste disposal or recycling. Therefore, energy supply indicators should be identified in the impact assessments of each step of the ESC to analyze and evaluate the present and future impacts of energy consumption on the three significant dimensions of sustainable development: environmental, social, and economic. The 30 energy indicators for sustainable development (EISDs) are categorized along these three dimensions. Well-analyzed and interpreted indicators act as an essential tool for communicating energy and sustainable development data to policymakers and to the public, give a better insight of the whole system including interlinkages between dimensions, and quantify progress toward sustainable development by monitoring the results of past policies (Vera & Abdalla, 2006).

The economic dimension shows the use and production trends of energy as well as energy supply security; the social aspect reflects the consequences caused by energy services on social well-being, focusing on equity and health; the environmental dimension determines the effects of the energy system on the environment, especially atmospheric, land, and water quality. The list of EISDs is given in Fig. 4.5. The EISDs are designed for all types of energy sources; hence, some energy indicators are selected to apply to RE supply. There is no final or definitive set of energy indicators, so indicators must be evaluated over time to suit and satisfy a country's conditions, resources, and capabilities.

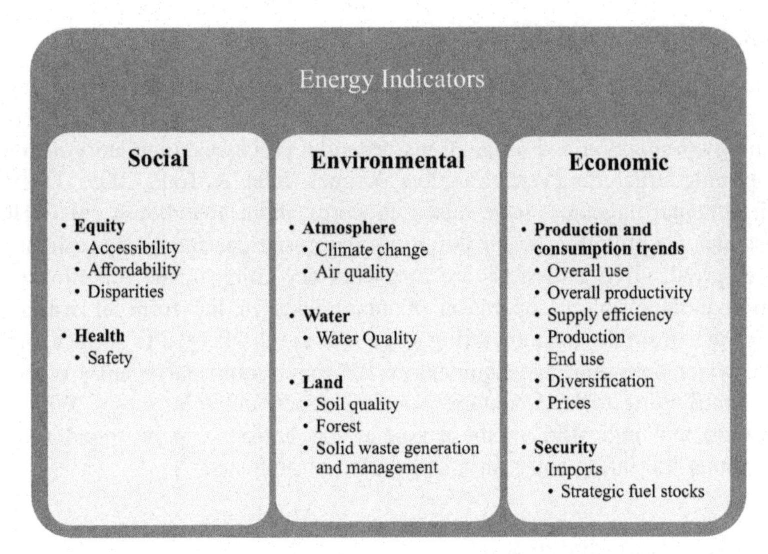

FIGURE 4.5 Indicators for sustainable development (IAEA, 2005).

4.5 Sustainability indicators of renewable energy technologies

Evaluation and analysis of RE projects require a combination of quantitative analysis and qualitative analysis. In line with the needs of social development, multiple indicators are selected from the three aspects of environment, society, and economy for comprehensive evaluation and analysis.

4.5.1 Social dimension

RE is inexhaustible energy with a variety of impacts that can be either positive or harmful to society. The key indicators in the social dimension are accessibility, affordability, disparities, and safety. Many households in rural areas still are not connected to the electricity grid and are therefore unable to access energy supplies. Lack of energy supply creates limitations and difficulties in life, as works must be carried out in a conventional way which is inefficient and time-consuming. Renewable energy supply should be distributed evenly all around the world and make it easily accessible. The social equity indicator takes degree of fairness into account to ensure affordability in the deployment of RE. RE should be made affordable, not just for the upper class but also the low and middle classes. Disparities between regions and income groups should be considered. Unequal distribution of income, energy transport, and distribution networks can cause differences within a country or between countries around the globe (IAEA, 2005). The safety indicator, which shows accident fatalities per energy produced, is evaluated at each stage of the ESC.

4.5.2 Environmental dimension

In the environmental dimension, energy indicators measure effects to air, land, and water quality. Ecological impacts may change depending on the methods of energy production and usage, transformation processes, regulatory actions, and pricing structures (Vera, Langlois, Rogner, Jalal, & Toth, 2005). Overall environmental indicators have raised concerns about greenhouse gas (GHG) emissions, air pollution, water pollution, deforestation, and waste generation and disposal. GHG emissions are measured according to the full life cycle from manufacturing to operation. Contaminants in the form of wasts or wastewater from the manufacturing stage that are discharged to surface water cause water pollution. Development of RE may require large areas of land, thus contributing to land contamination by producing solid waste. With the environmental indicator, all these potential pollutions can be monitored to ensure that the safety and health of people are protected.

4.5.3 Economic dimension

Two main concerns in the economic dimension of energy supply are production utilization and trend and security of energy supply. In overall use, energy use per capita expresses the amount of energy use per person in a given country or area (Vera et al., 2005). In overall productivity, energy use by GDP directly indicates energy intensity. High cost of transforming energy into GDP is indicated by high energy intensity factor. The efficiency of energy conversion and distribution should be monitored. High supply efficiency provides cost-effective transformation processes. Production indicator gives the data of energy consumption per resources used. There are different types of energy indicators of energy intensities for various sectors, such as industry, agriculture, service/commercial, household and transport for RE supply. The purpose of energy diversification is to decrease the reliance and dependence on a single resource. Diversifying energy mix can prevent energy disruptions, strengthen energy security and reduce the risks posed by political unrest or natural disasters (Hanania, Stenhouse, & Donev, 2018).

The high price of energy supply is one of the barriers to electricity accessibility and affordability. This can be directly affected by energy taxes and subsidies. Energy must be available at all times and have net energy imports dependency. Energy security is essential to economic and financial development as energy supply interruptions can cause significant losses to a country.

4.5.3.1 Energy security

In the ESC right from production to final end use, energy security should be considered as there are certain specific risks along this ESC. There are four resource indicators to examine the risk associated with imported resources, which are:

(a) import dependency (ID),
(b) RE sources on primary energy (PE_{RES}),
(c) supplier concentration through the Herfindahl–Hirschman index (HHI) (HHI_{global} and HHI_{sup}), and
(d) energy bill (Gouveia, Dias, Martins, & Seixas, 2014).

These indicators can be calculated as shown in Table 4.1. HHI is an index that measures the market concentration of a certain industry. It is used to monitor the potential impact of mergers and acquisitions on an industry (CFIE, 2018).

4.5.3.2 Energy dependency

Import dependency (ID) is the aspect of energy supply security. The diversity of energy resources is essential in the development of a sustainable energy future. It has been observed that total energy consumption in Malaysia is consistently increasing since the year 1990. Malaysia presents a lower degree of dependence on fuel imports. ID trend of Malaysia has dropped since 2015; this might occur due to the growth of RE in Malaysia, which in turn led to a drop in dependency on fossil fuels.

Import dependency (ID) can be calculated as:

$$ID = \left[\sum (fimp/TEC)\right] \times 100\%$$

Malaysia seems to be more dependent on imported fossil fuel energy resources with ID of 47% for year 2015.

TABLE 4.1 Indicators to examine the risk (Gouveia et al., 2014).

Import dependency	Renewable energy sources on primary energy (PE_{RES})	Supplier concentration through HHI
Important dependency (ID) = [Σ (fimp)/total energy consumption (TEC)] x 100% where f_{imp} is the import of fuel and TEC is total energy consumption to generate electricity plus net imports.	$PE_{RES} = RES/TPE_{elc}$ where RES is renewable energy sources and TPE_{elc} is total primary energy consumption for the power sector.	$HHI_{sup} = \Sigma S_{if}^2$ where HHI_{sup} for fuel and S_i is a share of each supplier of fuel.

4.5.3.3 Energy supply diversity

It is vital to reduce reliance on environmentally damaging and expensive fossil fuels. Two indices, the Shannon–Wiener index (SWI) and HHI, respectively are used to assess diversity and concentration (Chalvatzis & Ioannidis, 2017). The main drivers for the growth of diversity are reductions in transport fuel and increases in renewable energy sources (RES). Apart from ID, diversity of energy resources is also important to energy supply security. In this research, the two indices used to assess the diversity and concentration of fossil fuel energy supply are SWI and HHI. It is vital to note that HHI is a measure of concentration and is the exact opposite of the diversity measure represented by SWI. Hence, their trends will appear to mirror each other (Chalvatzis & Ioannidis, 2017). Over the years, technology has pushed RE sectors to be more and more efficient to satisfy energy demands. Each type of renewable and non-RE sources has the potential to significantly contribute to energy production.

4.6 Energy supply forecasting

Worldwide energy consumption is expected to grow 37% by the year 2040, with energy consumption plateaued in much of Europe, Korea, Japan, and North America. Also, growth in consumption is concentrated in the rest of Asia (60% of the global total), the Middle East, sub-Saharan Africa, and Latin America.

4.6.1 Renewable energy forecasting methodology

RE forecasting in its nascent stage began in the early years of the 21st century. Because of the large amount of uncertainty regarding RE, methods developed with a large, disparate perspective. It is quite difficult to categorize the existing methodology for forecasting because each model forecasts the certain parameters of installed and generation capacity, cost of generation, demand and supply, etc. There are various methods employed to forecast RE which includes analyzing the status or future perspective predicting with a focus on specific target of interest. Most forecasting simulations studies broadly present the existing landmark and predictions of a target country or region and recommends policy and schemes for future penetration. These studies work well on developing countries whereby presenting a wide-ranging status of RE in that nation-state and recommend strategies custom tailored for that country.

One of the most widely used RE forecasting models is the energy technology perspectives—the integrated MARKAL-EFOM system (ETP-TIMES). This model works on combination of several points connected to energy demand and supply, it then computes recommendations based on four consolidated models i.e., energy conversion, buildings, industry and transportation. This model can perform detailed analysis of energy infrastructure and various technologies in form of scenario particularly it offers in the field of power generation. However, the model has a few limitations in reflecting changes in the energy technology costs and noneconomic characteristics of each technology that can have severe impacts on future technology competitiveness. There are two approaches of forecasting models— i.e., a top-down and bottom-up approach. These two approaches have their own pros and cons. A forecast is made at the uppermost level and then is proportioned accordingly in a top-down approach, whereas in a bottom-up approach, lower-level forecasts are accumulated to higher levels in the forecasting hierarchy. Fig. 4.6 illustrates an example of a bottomd-up approach method encompassing various forecasting models, such as the competitive diffusion model, linear regression model, logistics growth model, government energy policies, and companies' projected intentions.

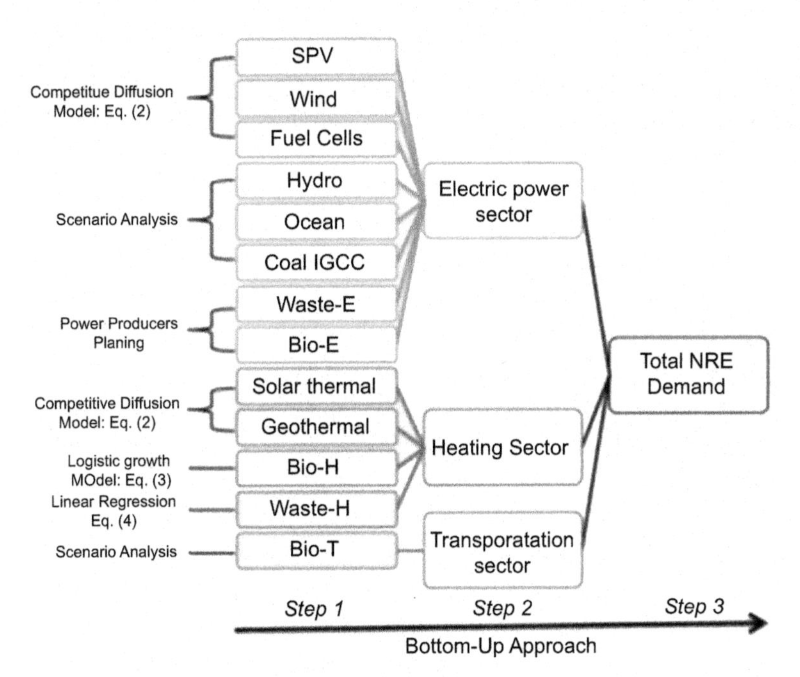

FIGURE 4.6 Bottoms-up approach for energy forecasting (Lee & Huh, 2017).

4.6.2 Renewable energy forecasting: global overview

Renewable energy (RE) has continued to make strides in global total final energy consumption (TFEC), accounting for more than 10.4% in 2016. The lion's share of the modern renewable share was renewable electricity, making up 5.7%, of which 3.7% was generated by hydropower. This is followed by renewable thermal energy (4.1%) and transport biofuels (0.9%). In addition to this, biomass consumption, primarily for cooking and space heating, made up an additional 7.8%. Therefore, approximately 18.2% is the total RE accounted in TFEC (Fig. 4.7) (IRENA, 2018). The overall contribution of RE to overall energy consumption has advanced only slightly due to the continued rise of overall energy demand, with omission in the year 2009 following the global economic recession.

The renewable heating, cooling, and transport sectors continue to mark infinitesimally slow progress despite government policies and steps to enhance the role and share of RE for heating and transport in the existing system. From the past 10 years, these two areas of RE have been recognized as the "sleeping giant of renewable energy potential" and continuously sidelined by lawmakers and policymakers in comparison to RE generation. The supply of heat via RE increased approximately 20.5% from 2007 to 2015; however, electricity supply from RE increased by approximately 56.6% during this same period. In areas of transport, the vast sum of most global energy demands is still fulfilled by biofuels and electricity by 2.8% and 1.3%, respectively. Bioenergy remained the front-runner up to 2014 in the contribution of RE to transport and heat supply (Fig. 4.8).

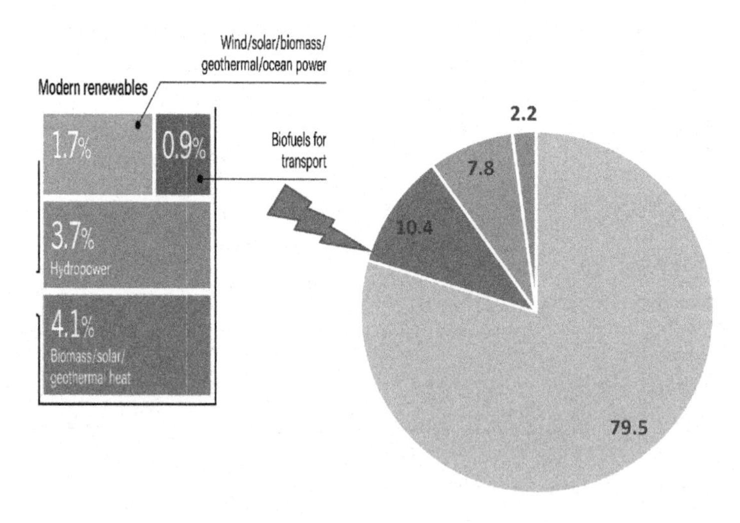

FIGURE 4.7 Estimated renewable share of total final energy consumption in 2016 (IRENA, 2018).

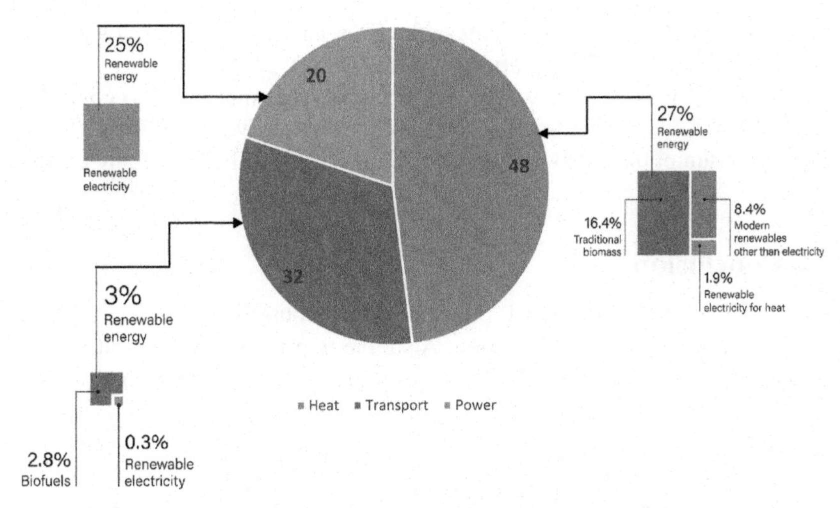

FIGURE 4.8 Sectorial renewable energy in total final consumption (IRENA, 2018).

4.6.3 Renewable energy forecasting: 2016–40

This forecasting extends into a long-term projection between the years 2016–2040. Various projections have been performed by multiple organizations, namely energy companies and institutions such as BP, CNPC, Statoil, OPEC, and EIA. All outlooks in the forecasts display continuous growth in global energy demand of 0.9%–1.4% per annum (Fig. 4.9). In this energy transition scenario, oil and gas grow by a combined rate equaling an average of other

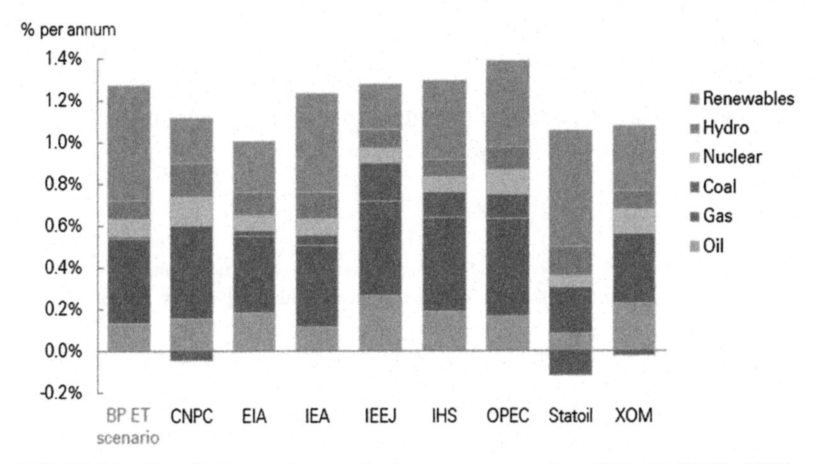

FIGURE 4.9 Contributions to the growth of energy consumption, 2016–40 (BPEO, 2018).

outlooks. The growth of renewables in the ET scenario (7% p.a.) is toward the top end of the sample of the various visions (BPEO, 2018).

The continuous stride toward a lower carbon fuel mix is set to continue, with RE and hydro climbing to contribute a cumulative of 40% of overall energy consumption by 2040, the most significant contribution of any energy source.

4.7 Conclusion

Renewable energy has broad prospects, and countries have made good investments in RE power generation. Also, the transportation and heating and cooling fields have great potential: transportation is the world's second-largest energy terminal consumption sector, accounting for about 29% of final energy consumption. And the use of energy by heating and cooling accounts for nearly half of all terminal energy consumption, wherease RE in heating and cooling face more obstacles than power generation and transportation. Through research on the RESC, with encouragement via government policies to achieve perfect usage of RE in different fields.

In terms of energy security, evaluation made using the three indicators shows general energy security performance. By this assessment, the diversity of energy sources increases by introducing more renewable sources in power generation. Although Malaysia is a relatively a small country and not immensely developed compare with other nations of the world, it has a more diverse fuel mix. Malaysia's net import is expected to increase over the years, as demand for fossil fuels is growing while the fossil fuel reserves are diminishing. Generally, the growth of oil import dependence in Malaysia is inevitable, as transportation and industrial sectors in Malaysia are still greatly dependent on oil products, and this puts the energy security of the country at risk.

It is essential to aim for alternative energy to reduce energy security risk as fossil fuels are being depleted, and also for the sake of the environment by reducing the carbon emissions resulting from releases from fossil fuel power plants. Renewable energy resources play an essential role in decreasing import dependence and at the same time contributing to diversification in the fuel mix, even if exposed to intermittency.

References

BPEO. (2018). *BP Energy Outlook 2018 edition.* https://www.bp.com/content/dam/bp/business-sites/en/global/corporate/pdfs/energy-economics/energy-outlook/bp-energy-outlook-2018.pdf (11/07/2018).

CFIE. (2018). *CFI Education Inc. (2018). Herfindahl-Hirschman Index (HHI).* Retrieved from Corporate Finance Institute https://corporatefinanceinstitute.com/resources/knowledge/finance/herfindahl-hirschman-index-hhi/.

Chalvatzis, K. J., & Ioannidis, A. (2017). Energy supply security in southern Europe and Ireland. *Energy Procedia, 105*, 2916–2922. https://doi.org/10.1016/j.egypro.2017.03.660.

Evans, A., Strezov, V., & Evans, T. J. (2009). Assessment of sustainability indicators for renewable energy technologies. *Renewable and Sustainable Energy Reviews, 13*(5), 1082–1088. https://doi.org/10.1016/j.rser.2008.03.008.

Feng, L., Mears, L., Beaufort, C., & Schulte, J. (2016). Energy, economy, and environment analysis and optimization on manufacturing plant energy supply system. *Energy Conversion and Management, 117*, 454–465. https://doi.org/10.1016/j.enconman.2016.03.031.

Gouveia, J. P., Dias, L., Martins, I., & Seixas, J. (2014). Effects of renewables penetration on the security of Portuguese electricity supply. *Applied Energy, 123*, 438–447. https://doi.org/10.1016/j.apenergy.2014.01.038.

Hanania, J., Stenhouse, K., & Donev, J. (2018). *Energy use per person*. Retrieved from Energy Education https://energyeducation.ca/encyclopedia/Energy_use_per_person.

Han, S., & Kim, J. (2017). An optimization model to design and analysis of renewable energy supply strategies for residential sector. *Renewable Energy, 112*, 222–234. https://doi.org/10.1016/j.renene.2017.05.030.

IAEA. (2005). *International atomic energy agency. (2005). Energy indicators for sustainable development: Guidelines and methodologies*. Vienna: IAEA.

IRENA. (2018). *IRENA – international renewable energy agency. (2018). Off-grid for energy access*. Retrieved from International Renewable Energy Agency http://www.irena.org/offgrid.

Lee, C.-Y., & Huh, S.-Y. (2017). Forecasting new and renewable energy supply through a bottom-up approach: The case of South Korea. *Renewable and Sustainable Energy Reviews, 69*, 207–217. https://doi.org/10.1016/j.rser.2016.11.173.

Shen, Y.-C., Lin, G. T. R., Li, K.-P., & Yuan, B. J. C. (2010). An assessment of exploiting renewable energy sources with concerns of policy and technology. *Energy Policy, 38*(8), 4604–4616. https://doi.org/10.1016/j.enpol.2010.04.016.

UNDP. (2010). *United nations development programme, productive uses of renewable energy: Energy supply chain, united nations development programme*. Retrieved from http://www.undp.org/gef/pure/05/05_1/05_1_2.htmS.

Vera, I. A., & Abdalla, K. L. (2006). Energy indicators to assess sustainable development at the national level: Acting on the Johannesburg plan of implementation. *Energy Studies Review*, 154–169.

Vera, I., Langlois, L. M., Rogner, H., Jalal, A., & Toth, F. L. (2005). Indicators for sustainable energy development: An initiative by the international atomic energy agency. *Natural Resources Forum*, 274–283.

Xu, L., Pang, M., Zhang, L., Poganietz, W.-R., & Marathe, S. D. (2018). Life cycle assessment of onshore wind power systems in China. *Resources, Conservation and Recycling, 132*, 361–368. https://doi.org/10.1016/j.resconrec.2017.06.014.

Further reading

EIA. (2018a). *U.S. energy facts – energy explained, your guide to understanding energy – Energy Information Administration*. Eia.gov, 2018. [Online]. Available https://www.eia.gov/energyexplained/?page=us_energy_home.

EIA. (2018b). *U.S. Energy Information Administration, "primary energy Overview"*. Washington DC: U.S. EIA.

Ilgin, M. A., & Gupta, S. M. (2010). Environmentally conscious manufacturing and product recovery (ECMPRO): A review of the state of the art. *Journal of Environmental Management, 91*(3), 563–591. https://doi.org/10.1016/j.jenvman.2009.09.037.

MESH. (2017). *Malaysia energy statistic handbook 2017, suruhanjaya tenaga (energy Commission), Putrajaya, Malaysia, 2018.*

MHCRC. (2018). *Malaysia's hunger for coal raises concerns – the Malaysian Reserve* [Online]. Available https://themalaysianreserve.com/2017/03/31/malaysias-hunger-for-coal-raises- concerns/.

Ream, M. K. (2015). *When it comes to energy, countries should mix it up*. Retrieved from Share America https://share.america.gov/diversifying-energy-sources-boosts-security/.

Chapter 5

Energy demand forecasting

M.A. Islam, Hang Seng Che, M. Hasanuzzaman, N.A. Rahim
Higher Institution Centre of Excellence (HICoE), UM Power Energy Dedicated Advanced Centre (UMPEDAC), Level 4, Wisma R&D, University of Malaya, Jalan Pantai Baharu, Kuala Lumpur, Malaysia

5.1 Introduction

The process of prediction, estimation, or projection of future activities, occurrences, or events is called forecasting. Hence, energy demand forecasting is the estimating of future energy consumption based on the various data and information available. Generally, demand forecasting is done by projecting past data/information into some mathematical models to predict the trend of future energy demand. This includes no just the magnitude but also location of load demand as a function of time. Such information is essential for financial and operational planning in electricity distribution systems.

Generally, energy demand forecasting can be classified into three categories based on the forecasting horizon,—i.e., how far into the future is the prediction done?

Types of forecasting	Forecasting horizon	Purpose
Short-term	1 h to 1 week	Normally applied for scheduling and analyses of the distribution network
Medium-term	1 month to 5 years	Normally used for planning resources and tariffs,
Long-term	5–20 years	Managing resources and making development plan

Accurate forecasting of energy demand is a very important yet challenging task because the change of energy demand with time can be affected by various environmental and socioeconomic factors (Abdel-Aal, 2008). Fig. 5.1 shows

Energy for Sustainable Development. https://doi.org/10.1016/B978-0-12-814645-3.00005-5

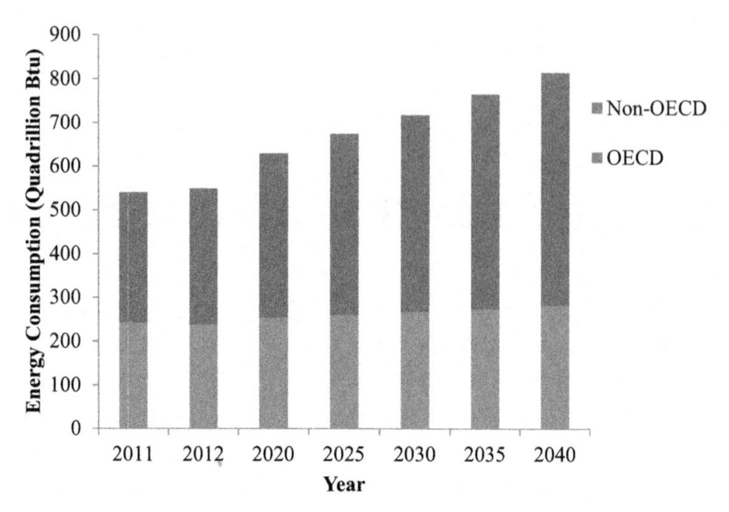

FIGURE 5.1 World energy consumption history (2011–12) and future energy demand forecasting to the year 2040 (EIA, 2016).

world energy consumption history (2011–2012) and future energy demand forecasting to the year 2040 by the International Energy Agency.

5.2 Importance of energy demand forecasting

Energy demand forecasting is essential for initiating development plans, making operational and maintenance decision, and performing strategic energy planning. The accuracy of energy demand forecasting can affect the costs and revenues for electrical power producers and transmission/distribution operators and hence the profitability and sustainability of these organizations. For policy makers, forecasted energy demand is useful for making policy and plans related to energy security and infrastructure development. The importance of energy demand forecasting includes (Agrawal, 2014)

- policy and planning—enables policy makers to make good planning decisions based on forecast future energy demand
- optimum supply schedule—allows power producers to dispatch power optimally, reducing risks of under- or overgeneration by the utility company
- fuel mix selection—facilitates the planning of fuel supplies and selection of fuel mix for power plants
- power plant and network planning—important in deciding on the size, location/route, nd types of future power plants and power transmission/ distribution networks
- demand-side management—allows utilities or policy makers to structure plans for managing demand through incentives, penalties, and other demand management measures
- renewable planning— facilitates decisions on the location and size of renewable energy generation plants

5.3 Challenges in energy demand forecasting

Energy demand can be affected by socioeconomic and environmental factors, particularly climatic conditions. For example, during winter, a temperature reduction of 1°C causes an additional 1.8 GW of power demand in France (Fischer, 2008).

Accurate energy demand forecasting hence relies on the assumptions for weather condition. Sudden and unpredicted climate changes can significantly affect the energy demand prediction, particularly for short- and medium-term predictions. In addition, geological differences in weather patterns should be considered when forecasting electricity demand.

Some other technical challenges are as follows:

- Uncertainty in energy consumption behavior due to changes in economic and social factors. For instance, sudden increases in electricity prices or a strike in factories might drive the consumption down.
- Difficulties due to the complex nature of loads.
- Difficulties in deriving accurate correlations between energy demand and influencing factors. This affects the mathematical model from which the demand forecast is made.

5.4 Forecasting process

A typical energy demand forecasting process is shown in Fig. 5.2. The forecasting procedure commonly starts from the energy database, which is essentially historical data that are reliable and accessible. This database should include the energy demand data as well as other variables that can affect the

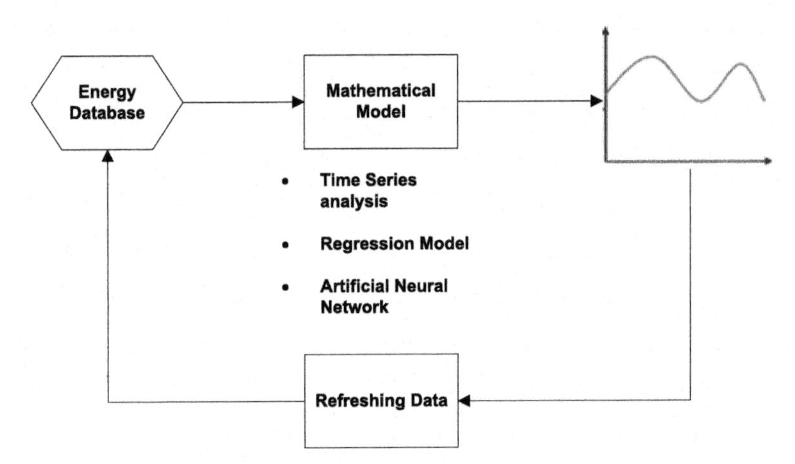

FIGURE 5.2 Forecasting method (Schellong, 2011).

energy consumption. The data is used as the inputs for a forecast model, whose output will be the forecasted energy demand. This is usually a continuous process, where new data will be added to refresh the database and new forecast is made as time progress.

There is a good variety of demand forecasting methods available in the literature. Regardless of the approach, following basic steps are usually required:

- **Setting the scope of the forecasted variable(s):**

It is necessary to first decide the scope of the forecasted variable(s). This includes: the detailed items to be forecasted, the units of the variable(s) as well as the forecasting horizon (Reid & Sanders, 2011). In general, this step tries to answer the What, Where and When questions.

What—For instance, energy demand forecast can be done either for overall energy consumption of for specific energy sources (oil, gas, electricity, etc.).

Where—The physical/geographical boundary of the variable should also be decided here. For example, the forecasted energy demand can be considered for a building, a district, a country, a region or even the world.

When—Finally the forecast horizon, i.e., short-, medium- or long-term, should be decided to serve the purpose of the forecast.

- **Evaluating the data:**

Once the scope of forecast is decided, it is then necessary to evaluate the data for determine what data is required and if these data is available. The required data and its availability will determine the methodology and accuracy of the forecast accuracy.

- **Select a forecasting model:**

With the data evaluation done, a suitable forecasting model needs to be selected next. The choice of model is usually tradeoffs between accuracy, complexity, availability of data as well as cost. More discussions on the commonly forecasting models will be given in the subsequent sections of this chapter.

- **Perform forecasting:**

With the database and forecasting model ready, this step basically involves inputting the data into the model to get the forecasted results.

- **Monitor forecast accuracy:**

Demand forecasting is usually a continuing process; though one-off forecast is also used for one-off projects. As discussed earlier, the accuracy of the forecasted demand can be affected by unpredicted socioeconomical and environmental factors. To continuously improve the forecast, it is required to monitor the forecast accuracy and make corresponding adjustments to the database and/or model.

5.5 Forecasting methods

The core of energy demand forecasting lies in the forecasting method used. To understand the forecasting methods, it is necessary to first understand the characteristics of energy demand time series. Energy demand time series are essentially energy demand data recorded at certain intervals of time. As a matter of fact, all demand prediction methods work around time series data, where the trend of previous energy demand data is analyzed in order to extrapolate future energy demand. Such time series analysis of energy demand data works on the fundamental assumption that the "pattern of the variable in the past will proceed into future".

In general, the pattern of variation for energy demand can be considered as a combination the following trend or patterns:

- long-term trend
- cyclical pattern
- seasonal pattern
- random or irregular pattern

In terms of long-term trend, it can be assumed that the energy demand will demonstrate either an increasing or declining trend with respect to time as a consequence of the overall change in political and socioeconomical factors etc. Fig. 5.3 shows a typical variation of energy demand in long-term trend pattern, where linear increase can be observed.

Instead of a monotonic increase of decline, it is known that the energy demand changes in cyclic manner over a long duration, usually many years of even decades. Such cyclic change in energy demand is commonly associated with the cyclic nature of economy. On a shorter timeframe, cyclic change in energy demand can also be observed on a daily or weekly basis, where energy increases and decreases during working and nonworking hours, as well as during weekdays and weekends respectively. Fig. 5 4 shows an example of energy demand curve with cyclical trend.

Fig. 5.5 shows the energy demand with seasonal trend. Such energy demand variations are the consequences of changes in human activities and

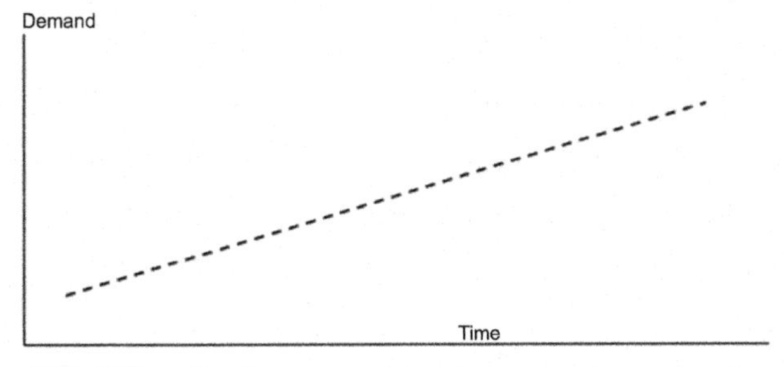

FIGURE 5.3 Trend approach of time series energy demand model.

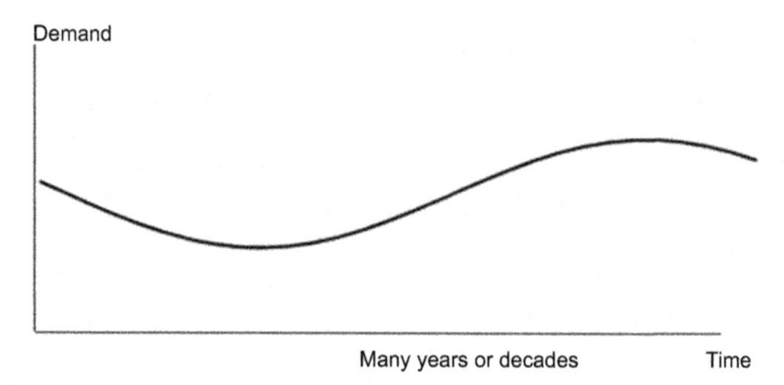

FIGURE 5.4 Cycle mode of time series energy demand model.

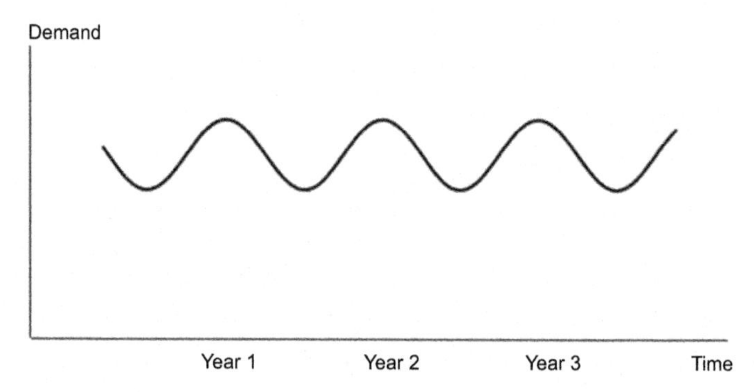

FIGURE 5.5 Seasonal mode of time series energy demand model.

equipment use due to different seasons in the year. The seasonal change can be attributed to the change of weather, where energy demand will increase or decrease with different space cooling and heating requirement for different seasons. Of course, the seasonal trend can also be due to human activities change, such as cultural or religious festivals as well as holidays. Unlike the long-term trend, the seasonal trend is observed over a shorter period, usually in the duration of several days, weeks or months.

Apart from periodic changes that exhibit repetitive patterns, random changes of energy demand are also commonly observed. Such random change is unpredictable and does not follow any fixed trend as shown in Fig. 5.6.

$Y(t) = T(t) + S(t) + C(t) + R(t)Y(t) = T(t) \times S(t) \times C(t) \times R(t)$ While the trends in energy demand be visually explained, such vague description is not useful to make reliable prediction. Forecasting of energy demand hence involves analyzing the history data, or time-series data, to find a model that best describe the data (i.e., time-series analysis) and subsequently using the model to make forecasts (time-series forecast). Since this requires the time-series analysis technique, the forecast method is also generally called time-series forecast in some literature (Deb, Zhang, Yang, Lee, & Shah, 2017).

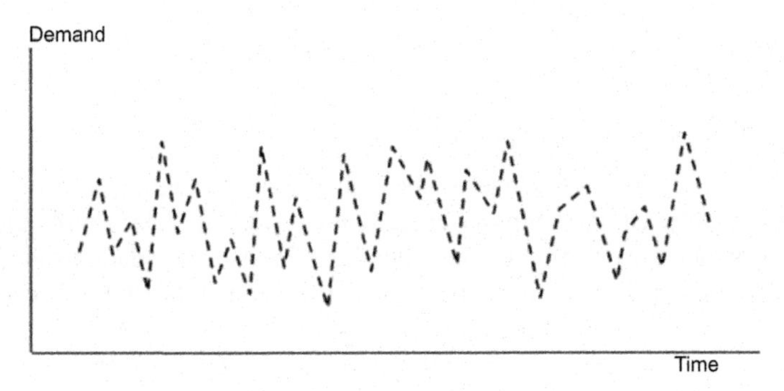

FIGURE 5.6 Random mode of time series energy demand model.

Regardless the method used, there are always two elements in time series analysis of energy demand:

(1) First, the identification of the underlying pattern or structure, i.e., defining a suitable model that best describe the data trend. Different types of mathematical models can be used for the curve fitting such as linear, logarithmic, polynomial, exponential etc.
(2) Second, performing curve fitting in order to obtain a best fit between the data and the model.

The two elements can be explicit, where the model and curve fitting methods can be separately selected or integrated, such as in averaging or artificial intelligence-based approaches where the model and curve-fitting are inseparable.

There are different ways of classifying the energy demand forecasting methods, either based on the model such as using static versus dynamic model, empirical versus mathematical model, univariate versus multivariate model, etc., or based on the curve fitting method statistical versus artificial intelligence method (Suganthi & Samuel, 2012).

Here, discussions on energy demand forecasting method are done based on the following categories:

- Averaging Method
- Mathematical Model based method
  - Regression Method
  - Autoregressive Method
- Artificial intelligence method

5.5.1 Averaging method

The simplest form of energy demand forecast is to extrapolate the time series data using averaging methods. For this method, Different types of averaging methods can be used for time series forecasting, as described as follows (Reid & Sanders, 2011);

5.5.1.1 Simple moving average

In this method, forecast is done by averaging a given number of most recent data repetitively. Unlike simple average that consider all data, simple moving average method only consider a fixed number of most recent data, by dropping old data as new data is acquired. In essence, the data points used by the simple moving average "moves" over time.

The simple moving average method (Reid & Sanders, 2011)

$$F_{t+1} = \frac{\sum E_t}{n} = \frac{E_t + E_{t-1} + \cdots + E_{t-n}}{n} \tag{5.1}$$

where

F_{t+1} is forecasted value of demand for the time $t+1$
E_t is real value for current period, t
n is number of data used for the moving average

5.5.1.2 Exponential smoothing

Exponential smoothing or exponential moving average is similar to weight moving average with two unique exceptions:

(i) Instead of putting weights on fixed number of data, exponential smoothing method put weights on all data points

(ii) The values of weight is selected to be exponentially decreases from more recent data to the oldest data.

The mathematical expression of exponential smoothing method of forecasting is as shown below (Reid & Sanders, 2011):

$$F_{n+1} = \alpha E_n + (1-\alpha)F_n F_{n+1} = \alpha E_n + (1-\alpha)E_n \tag{5.2}$$

where

F_{n+1} is forecasted value of demand for the period $(n + 1)$
E_n is real value of energy demand for the current period, n
F_n is forecasted value of demand for the period, n
α is the coefficient for smoothing (value of α is between 0 and 1).

The best choice of α can be determined by trial and error, but the general guideline for α selection is

$$\alpha = \frac{2}{n+1}$$

where n is the number of periods.

Even though averaging method is very simple and fast, such methods are unable to fully describe the relation between energy demand and other influencing factors. As a matter of fact, the averaging method inherently assumes that the energy demand is purely a function of time. It is not possible to simultaneously consider the effect of time and other factors on energy demand using averaging. Hence, mathematical model approaches and artificial intelligence methods are more popular to increase the accuracy of the forecast. These are discussed in the subsequent part of this chapter.

5.5.2 Regression methods

Regression is one of the most widely used statistical techniques for forecasting, including energy demand forecast. The energy demand forecast using regression method consists of two steps:

(1) To construct a regression model to expresses a general relationship between energy demand (known as the response/dependent variable or the variate) and variables that can affect the demand (known as forecaster/independent variable), i.e., such as weather, consumer's behavior, number of working hours./days etc.
(2) To perform regression analysis, an iterative process so that repetitively analyze, verify, criticize, and modify the coefficients in the model until a good relation between the dependent variables and the independent variables can be determined.

Classification of regression method is as follows (Fumo & Rafe Biswas, 2015; Sobri, Koohi-Kamali, & Rahim, 2018):

1. Linear regression
   - **(a)** Univariate
     - **(i)** Simple linear regression
     - **(ii)** Multiple linear regression
   - **(b)** Multi Variate regression
2. Nonlinear regression

5.6 Simple and multiple linear regression

The linear regression analysis is a linear approach for modeling the relationship between the dependent variables and one or more independent variables. If there are more than one independent variables then the regression is called as multiple linear regression. It ought to be comprehended that a relationship among a response variable and an indicator variable does not really suggest that the indicator variable causes the response variable, however, that there is some critical relationship between the two variables. According to the complexity of relationship among the variables, the linear regression can be classified into main types: (1) Simple linear regression and (2) multiple linear regression.

Regression analysis is called as “univariate regression” when only one dependent parameter is involved and called “multivariate regression” where two or more dependent parameters need to be forecasted.

5.6.1 Simple linear regression

The mathematical expression of simple linear regression is as follows (Fumo & Rafe Biswas, 2015; Sobri et al., 2018):

$$Y = \beta_0 + \beta_1 X + \varepsilon \tag{5.3}$$

where

Y is the dependent or response variable;
X is the forecaster or independent variable;
β_0 and β_1 are the coefficients of regression;
ε refers to error to calculate the difference between forecasted and observed data.

The β_0 value can be obtained from the intercept, that is the value of Y when the value of X is zero. β_1 is the slop of the line, rate of change of Y per unit change of X value. If $\widehat{\beta}_0$ and $\widehat{\beta}_1$ are the estimated value of the parameters β_0 and β_1 then $\widehat{Y}$ value, which is the fitted or forecasted value of Y, can be expressed as:

$$\widehat{Y} = \widehat{\beta}_0 + \widehat{\beta}_1 X \tag{5.4}$$

The least squares estimation procedure is used to the estimated value of the parameters β_0 and β_1 as follows:

$$\widehat{\beta}_0 = \frac{\sum (X - \overline{X})(Y - \overline{Y})}{\sum (X - \overline{X})^2} = \frac{\sum XY - n\overline{X}\,\overline{Y}}{\sum X^2 - n\overline{X}^2} \tag{5.5}$$

$$\widehat{\beta}_0 = \overline{Y} - \widehat{\beta}_1 \overline{X}$$

where

$\overline{Y}$ is mean of the Y values
$\overline{X}$ is mean of the X values
n is an observation or data number

5.6.2 Multiple linear regressions

The multiple linear regression contains more than one predictor variables. The mathematical expression of multiple linear regression is as follows (Fumo & Rafe Biswas, 2015; Sobri et al., 2018):

$$Y = \beta_0 + \beta_1 X_1 + \beta_2 X_2 + \cdots + \beta_p X_p + \varepsilon \tag{5.6}$$

where

Response parameter is Y
Predictor parameters ($X_1; X_2;\ldots$ Xp) with p being the number of parameters
Regression coefficients (β_0; β_1, g β_p) and
ε is the error to estimate the deviation of a forecasted value from the observed data.

The forecasted value form multiple linear regression is as follows:

$$\widehat{Y} = \widehat{\beta}_0 + \widehat{\beta}_1 X_1 + \widehat{\beta}_2 X_2 + \cdots + \widehat{\beta}_p X_p \tag{5.7}$$

where $\widehat{Y}$ is known as forecasted value and

$\widehat{\beta}$ is the regression coefficients need to estimate from the previous data.

5.7 Multivariate linear regression

Multivariate linear regression refers to the generalization of multiple linear regression methods. Multivariate linear regression examines the linear relationship probability between the multiple predictors and multiple response parameters. The following forecasting equation expresses the multivariate regression (Fumo & Rafe Biswas, 2015; Sobri et al., 2018):

$$\widehat{Y}_i = \widehat{\beta}_{i,0} + \widehat{\beta}_{i,1} X_1 + \widehat{\beta}_{i,2} X_2 + \cdots + \widehat{\beta}_{i,p} X_p \tag{5.8}$$

In a multiple regression process the following correlations can be obtained:

- Correlations between the predictor variables and response variables
- Correlations between the different response variables

For example, the relationship between the household electricity consumption and different environmental parameters such as ambient temperature, humidity, solar radiation etc. These three indicator factors correspond with each other. If the power utilization is integrated into HVAC systems, lighting, and machines, these dependent variables are also associated with each other.

5.8 Nonlinear regression

Energy demand forecasting by a nonlinear regression method contains following characteristics:

- Observed data are modeled by a nonlinear function,
- Response variables depend on one or more independent variables,
- Like linear regression method, nonlinear regression methods also include single or multi-variable regression.

The nonlinear regression can be expressed by general mathematical function (f) that describes the relationship between the response or dependent variables and predictors or independent variables as follows:

$$Y = f(X, \beta) + \varepsilon \tag{5.9}$$

where β is the unknown parameters (regression coefficients) and ε is an error term. The forecasting equation is given as:

$$\widehat{Y} = f\left(X, \widehat{\beta}\right) \tag{5.10}$$

5.8.1 Autoregressive model

The autoregressive (AR) method offers a different modeling method requiring only data on the previous modeled variable. In the AR process, the current value of energy demand is often expressed as a linear combination of previous actual values and with a random noise. The name *auto*regressive means self-regression (the Greek prefix *auto* means "self"). The process is basically a linear regression of the data in the current series against one or more past values in the same series. Mathematically, AR model can be defined as (Sobri et al., 2018):

$$Y(t) = \delta + \phi_1 Y_{t-1} + \phi_2 Y_{t-2} + \cdots + \phi_p Y_{t-p} + a_t \tag{5.11}$$

where

Y_{t-1}, Y_{t-2}, ..., Y_{t-p} are the past series values (lags),
a_t is white noise (i.e., randomness), φ is regression coefficient
δ is a constant defined by the equation as: $\delta = \left(1 - \sum_{i=1}^{p} \phi\right)\mu$
μ is the process mean
p is called the order of the auto regression

By combining two models, such as AR and moving average (MA), a new autoregressive integrated moving average (ARMA) model is formed. ARMA is represented as (Sobri et al., 2018):

$$Y(t) = \sum_{i=1}^{p} \phi_i Y_{(t-1)} + \sum_{j=1}^{q} \beta_j e_{t-j} \tag{5.12}$$

where

$Y(t)$ represents value need to forecast prediction at time t.

For the AR part of the model (first term on the right-hand side of the equation):

p denotes the order of AR process,
φ_i is the ith coefficient of AR

For the MA part of the model (second term on the right-hand side of the equation):

q is the order of moving average error term,
β_j is the jth coefficient of MA
e_t is the white noise that refers uncorrelated random parameters at mean and constant of variance value zero

The main requirement of ARMA model is that it must be a stationary time-series.

Autoregressive integrated moving average (ARIMA) is an extension of ARMA model which is also widely used for predicting and modeling time-series. The ARIMA models can be considered a function of three variables, i.e., ARIMA (p,d,q), with "p" being the order of the AR part, "d" representing the degree of first differencing part and "q" representing the order of moving average (MA) portion. The details of equations 5.13–5.16 as follows (Sobri et al., 2018).

So the AR portion is

$$Y_{AR}(t) = \delta + \phi_1 Y_{t-1} + \phi_2 Y_{t-2} + \cdots + \phi_p Y_{t-p} + a_t \tag{5.13}$$

and the MA part is given as:

$$Y_{MA}(t) = \delta + \beta_1 e_{t-1} + \beta_2 e_{t-2} + \cdots + \beta_p e_{t-p} + a_t \tag{5.14}$$

while the ARMA can be expressed as:

$$\begin{aligned} Y(t) = \delta + \phi_1 Y_{t-1} + \phi_2 Y_{t-2} + \cdots + \phi_p Y_{t-p} + \beta_1 e_{t-1} \\ + \beta_2 e_{t-2} + \cdots + \beta_p e_{t-p} + a_t \end{aligned} \tag{5.15}$$

In addition, the difference ∇ is used to solve the nonstationary problem

$$\nabla^d Y(t) = \nabla^{d-1} Y_t - \nabla^{d-1} Y_{t-1} \tag{5.16}$$

Along with ARMA and ARIMA, there are some other similar models used for energy demand forecasting such as:

ARMAX = Autoregressive moving average with exogenous variable
ARIMAX = Autoregressive integrated moving average with exogenous variable

ARMA models are typically utilized for stationary procedures while ARIMA is an augmentation of ARMA for nonstationary procedures.

5.8.2 Artificial neural networks

Apart from statistical methods discussed above, artificial intellig\eqalign{ent based forecasting methods have gained popularity due to their robustness against uncertainty. For the case of energy demand forecasting, the use of AI approach is especially helpful in situations where a suitable mathematical model of for the energy demand is not clear.

Among the different AI methods, the artificial neural network (ANN) method is most widely used for solving nonlinear and complex data without a prior assumption on the correlations between the data. Inspired by the human brain, the ANN model uses cascaded layers of nodes, known as neutrons, where the nodes between layers are correlated with one another using adjustable weights. By tuning the weights using a "training" process, the ANN can be conditioned as a black box model capable of providing input-output relation that matches the original training data. This allows ANN to be used extensively for modeling and signal processing, as well as for forecasting applications, including energy demand forecasting. The accuracy of ANN-based forecast depends on choice of ANN structure, choice of input parameters, training algorithm and sufficiency of training data. The complex nature of most practical problems can be effectively solved via ANN, due to ANN"s ability in. Fig. 5.7 shows the basic configuration of ANN that consists three layers known as input layers, hidden layers, and output layers.

To use ANN for forecasting, it is necessary to first construct a suitable ANN model. The process of ANN modeling involves three stages (as shown in Fig. 5.8)—i.e.:

(1) Design stage—the type of ANN, number of layers, number of neurons in each layer, input and output parameters, are decided during this stage;
(2) Training stage—an iterative process where historical data of the input and output parameters were presented to ANN and the weights that links between the neurons were modified until a predetermined qualifying condition was satisfied;

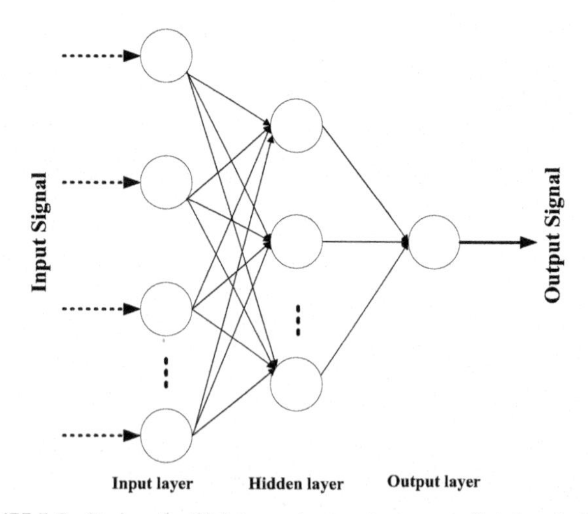

FIGURE 5.7 Basics of artificial neural network structure (Sobri et al., 2018).

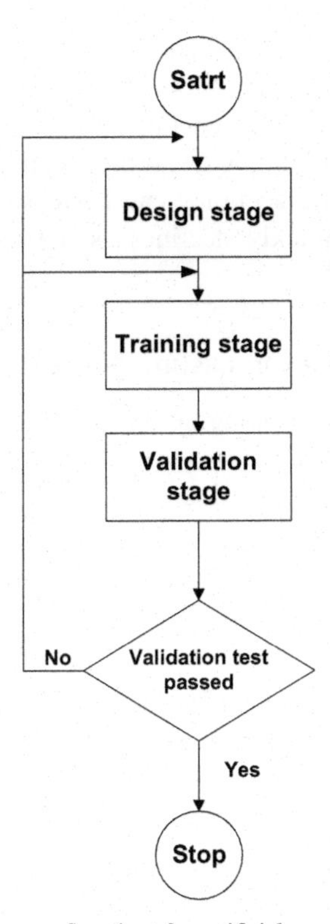

FIGURE 5.8 The modeling process flowchart for artificial neural networks (Sobri et al., 2018).

(3) Validation phase—the constructed ANN model is tested using data set that is different from those used in the training phase to see if the outputs can be predicted accurately using the input data. Once validation test is successful, the ANN model is ready for use.

5.9 Forecast accuracy

The accuracy of a forecasting method is one of the most important criteria in evaluating and selecting from the various available methods. It is worth noting that the accuracy of a forecasting model may degrades overtime due to various factors, such as the changes in data over time, increased or decreased influence of the independent variables on the dependent variables, as well as appearance of new variables that can affect the energy demand. Hence, the accuracy of forecasting models needs to be monitored periodically even after the forecasting method has been decided.

It is important to acknowledge that the forecasting accuracy is a combination of good data, good process, and a good model. The lack of or degradation of any aforementioned factors can have detrimental effect to the overall accuracy. Apart from that, more data does not necessarily result in better forecast. It is important to select optimum data size. In addition, interval of the independent variable data is also an important factor that affect forecasting accuracy. As shown in the figure below, if the independent data obtained are on hourly basis, the forecasting accuracy can be reasonably high within similar period (i.e., hours), but quickly declines as the interval increases to days, weeks and years (Fig. 5.9).

5.9.1 General metrics of forecasting accuracy

In order to evaluate the forecasting accuracy, the forecasting error has to be determined.The forecasting error is defined as the deviation of the forecasted value at time t from the actual value observed at time t, i.e., (Wallström & Segerstedt, 2010):

$$\varepsilon_t = E_t - F_t \tag{5.17}$$

where

ε_t is the error of forecasting for the period of t
E_t is the real value of energy demand that is observed after forecasting at time t
F_t is the forecasted energy demand at time t

The forecasted error can be positive or negative, with a lower absolute magnitude indicating higher forecasting accuracy and vice versa.

For the case of energy demand, the forecasting error can appear as a time-series rather than a single value. The forecasting error time series can have varying magnitude and polarity making a direct evaluation of forecasting accuracy difficult. In such case, some general metrices are commonly used to

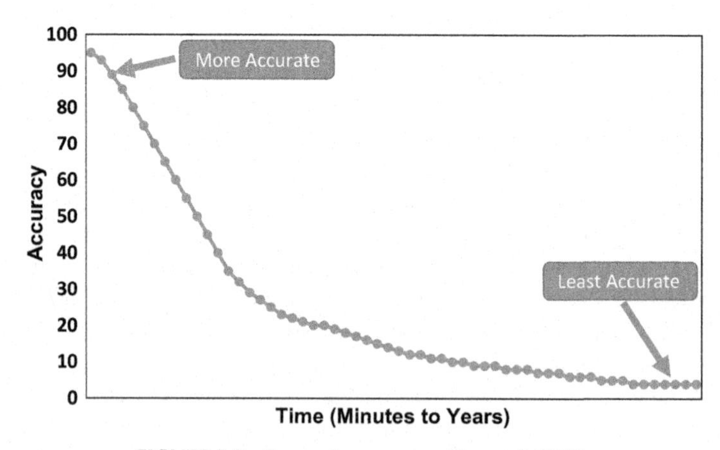

FIGURE 5.9 Forecasting accuracy (Agrawal, 2014).

evaluate the overall forecasting accuracy. The most widely used metrics are as follows:

- Mean absolute deviation (MAD)
- Mean squared error (MSE)
- Mean absolute percent error (MAPE)
- Symmetric mean absolute percent error (sMAPE)

5.10 Mean absolute deviation

MAD is an average of the absolute error. By using an absolute function, MAD considers the size of the error only, without considering the polarity. This ensures that errors with different polarities do not cancel one another and reducing the overall mean value. Hence for MAD, the lower value indicates the higher accuracy. The mathematical expression of MAD is as follows (Wallström & Segerstedt, 2010):

$$\mathrm{MAD} = \frac{\sum|\mathrm{Actual} - \mathrm{Forecast}|}{n} \quad \mathrm{MAD} = \frac{\sum|\mathrm{Error}|}{n} \tag{5.18}$$

For online forecasting, MAD can be continually updated by using following exponential smoothing technique as follows:

$$\mathbf{MAD}_t = \alpha(\boldsymbol{E}_t - \boldsymbol{F}_t) + (1 - \alpha)\mathbf{MAD}_{t-1} \tag{5.19}$$

where

MAD_t = Mean absolute deviation at period t
MAD_{t-1} = Mean absolute deviation at period $(t - 1)$
E_t = Actual value at period t
F_t = Forecasted value at period t
α = Smoothing constant

The forecasting is generally monitored by using tracking signal. The formula of tracking signal is as follows (Wallström & Segerstedt, 2010):

$$\mathrm{Tracking\_Signal} = \frac{\sum(E_t - F_t)}{\mathrm{MAD}_t} \tag{5.20}$$

where

MAD_t = Mean absolute deviation at period t
E_t = Actual value at period t
F_t = Forecasted value at period t

5.11 Mean square error

MSE is the average of the square of errors, which can be expressed as follows (Wallström & Segerstedt, 2010):

$$\text{MSE} = \frac{\sum (\text{Actual} - \text{Forecast})^2}{n} \quad (5.21)$$

Instead of using absolute function like in MAD, MSE uses the square of errors to eliminate the problem of the different polarity. In addition, data point with large errors are magnified due to the use of square function, giving a higher penalty for larger errors.

5.12 Mean absolute percent error

Instead of taking the error as an value, MAPE calculates the taking average of the absolute errors expressed this value as a percentage of the actual values, i.e (Sobri et al., 2018):

$$\text{MAPE} = \frac{100}{n} \sum_{t=1}^{n} \left| \frac{A_t - F_t}{A_t} \right| \quad (5.22)$$

where

n = number of observed data
A_t = Actual observed value
F_t = Forecasted value.

The advantage of MAPE is that the percentage of error with respect to the actual value is reflected. For example, the actual and forecasted energy demand for two data points are [1000 ktoe, 50 ktoe] and [1010 ktoe, 60 ktoe] respectively. While both data have the same absolute error of 10 ktoe, this error is only 1% of the first data, but 20% of the second data. Under situations where percentage error is important, MAPE will be useful. However, due to the division operation MAPE cannot be used in cases where the actual value is zero or close to zero (Tofallis, 2015).

5.13 Symmetric mean absolute percent error

SMAPE is a forecasting accuracy metric that overcomes the weakness of MAPE by allowing evaluation of accuracy while there is zero or near zero data. Similar to MAPE, sMAPE gives measurement based on percentage (or relative) errors, which can vary within a range of −200% to +200%. The mathematical formula of the sMAPE is given as follows (Wallström & Segerstedt, 2010):

$$\text{sMAPE} = \frac{100}{n} \sum_{t=1}^{n} \frac{|F_t - E_t|}{(|E_t| + |F_t|)/2} \quad (5.23)$$

where

n = number of observed data
E_t = actual observed value
F_t = forecasted value.

References

Abdel-Aal, R. E. (2008). Univariate modeling and forecasting of monthly energy demand time series using abductive and neural networks. *Computers and Industrial Engineering, 54*(4), 903–917. https://doi.org/10.1016/j.cie.2007.10.020.

Agrawal, A. (2014). Training program on "power procurement strategy and power exchanges", short term load forecasting. In *Indian Institute of Technology Kanpur (IITK) and Indian Energy Exchange (IEX), 28–30 July, 2014.*

Deb, C., Zhang, F., Yang, J., Lee, S. E., & Shah, K. W. (2017). A review on time series forecasting techniques for building energy consumption. *Renewable and Sustainable Energy Reviews, 74*(Suppl. C), 902–924. https://doi.org/10.1016/j.rser.2017.02.085.

EIA. (2016). *International energy outlook 2016.* Retrieved from https://www.eia.gov/outlooks/ieo/pdf/0484(2016).pdf.

Fischer, M. (2008). Modeling and forecasting energy demand: Principles and difficulties. In A. Troccoli (Ed.), *Management of weather and climate risk in the Energy Industry* (pp. 207–226). Dordrecht, Netherlands: Springer Verlag.

Fumo, N., & Rafe Biswas, M. A. (2015). Regression analysis for prediction of residential energy consumption. *Renewable and Sustainable Energy Reviews, 47*(Suppl. C), 332–343. https://doi.org/10.1016/j.rser.2015.03.035.

Reid, R. D., & Sanders, N. R. (2011). *Operations management: an integrated approach.* USA: John Wiley & Sons, Inc.

Schellong, W. (2011). Energy demand analysis and forecast. In G. Kini (Ed.), *Energy management systems.* InTech.

Sobri, S., Koohi-Kamali, S., & Rahim, N. A. (2018). Solar photovoltaic generation forecasting methods: A review. *Energy Conversion and Management, 156*, 459–497. https://doi.org/10.1016/j.enconman.2017.11.019.

Suganthi, L., & Samuel, A. A. (2012). Energy models for demand forecasting—a review. *Renewable and Sustainable Energy Reviews, 16*(2), 1223–1240. https://doi.org/10.1016/j.rser.2011.08.014.

Tofallis, C. (2015). A better measure of relative prediction accuracy for model selection and model estimation. *Journal of the Operational Research Society, 66*(8), 1352–1362. https://doi.org/10.1057/jors.2014.103.

Wallström, P., & Segerstedt, A. (2010). Evaluation of forecasting error measurements and techniques for intermittent demand. *International Journal of Production Economics, 128*(2), 625–636. https://doi.org/10.1016/j.ijpe.2010.07.013.

Chapter 6

Energy storage technologies

Fayaz Hussain[a], M. Zillur Rahman[b], Ashvini Nair Sivasengaran[a], M. Hasanuzzaman[a]
[a]*Higher Institution Centre of Excellence (HICoE), UM Power Energy Dedicated Advanced Centre (UMPEDAC), Level 4, Wisma R&D, University of Malaya, Jalan Pantai Baharu, Kuala Lumpur, Malaysia;* [b]*Department of Industrial Engineering, Faculty of Engineering, BGMEA University of Fashion and Technology (BUFT), Dhaka, Bangladesh*

6.1 Introduction

Global demand for energy is escalating rapidly with the snowballing of population size, which eventually will impact society and the environment (Hasanuzzaman et al., 2012). Renewable energy is the anticipated solution to this problem and is forecast to contribute about 2351 GW by April 2019 (Chen et al., 2009; Hasanuzzaman, Zubir, Ilham, & Che, 2017; Hedegaard & Meibom, 2012; Hossain, Hasanuzzaman, Rahim, & Ping, 2015; Jacob, 2001). Mismatched demand and supply is the major challenge for renewable energy sources due to their intermittent nature, which can be overcome by energy storage concepts (Leung et al., 2012). Energy storage technologies include electrochemical, pumped hydro, and compressed air, but batteries (regenerative fuel cell and rechargeable batteries) are the suitable and potential cost-effective options to sustain renewable energy systems to work effectively (Anne-Chloé Devic, 2018; Council, 2017; Leung et al., 2012; Li, 2016; Luo, Wang, Dooner, & Clarke, 2015; Park, Ryu, Wang, & Cho, 2017). The electricity storage objective is not limited to only peak shaving or short-term outage prevention these days, but rather there are many applications of EES such as voltage balancing (prevent burnouts), capacity/network expansion delay, and frequency regulation (Decourt & Debarre, 2013). Currently, energy storage integration has been proved to a beneficial option for the energy generation, transmission, and distribution sectors (Rastler, 2010; Strbac et al., 2012).

These are categorized based on time scale according to specific objective. Time scale categorization is done as instantaneous, short-term, midterm, and long-term depending on the time duration varying form seconds to days

Energy for Sustainable Development. https://doi.org/10.1016/B978-0-12-814645-3.00006-7

respectively (Akhil et al., 2013; Battke, Schmidt, Grosspietsch, & Hoffmann, 2013; Denholm et al., 2013; ENER, 2019; Hadjipaschalis, Poullikkas, & Efthimiou, 2009; Koohi-Kamali, Tyagi, Rahim, Panwar, & Mokhlis, 2013; Lemaire et al., 2011). Further, cost, size and operation of battery storage system is still under research and development (Hoffman, Kintner-Meyer, Sadovsky, & DeSteese, 2010). Battery energy storage for domestic purpose is rapidly increasing utilizing small units particularly where solar panels are installed (Khalilpour & Vassallo, 2016). Some systems are designed for domestic electrical energy management integrated with PV systems for the purpose of peak demand reduction and time of-use tariffs (Khalilpour, Vassallo, & Chapman, 2017; Lu & Shahidehpour, 2005). Safety of the battery energy storage is another serious concern as there have been incident of battery system failures when used at large scale, which released toxic materials to the environments. However, this technology needs further research and development in terms of efficiency, cost effectiveness and safety as it is promising and smart energy storage technology (Hazza, Pletcher, & Wills, 2004; Khalilpour et al., 2017; Lin, Chen et al., 2015; Liu et al., 2010; McKerracher, Ponce de Leon, Wills, Shah, & Walsh, 2015; Wei, Wu, Zhao, Zeng, & Ren, 2018). The literature of batteries is presented by classifying batteries by the materials used.

6.2 Battery energy storage technologies

Different types of battery energy storage technologies are developed and are used on commercial scale with characteristics from different aspects, selection of the type being dependent on the nature of its application such as charge-discharge rates, energy storage capacity, and power and response time. In addition, it is crucial to determine the design and construction of battery system in terms of its capacity and power.

6.2.1 Lithium-ion batteries

The lithium-ion (Li-ion) batteries functionality is based on the lithium ion movement between the positive and negative electrodes. Positive electrode releases the lithium ions to the negative electrode during charging mode through the electrolyte as medium of movement of ions. This phenomenon happens in opposite way when the battery is discharged. The material for the positive electrode is made of compounds made of lithium e.g., lithium manganese oxide and lithium iron phosphate-Li (Koenig, Belharouak, Deng, Sun, & Amine, 2016; Xu, Chen, Zhang, Zheng, & Li, 2015). However, graphite with sufficient specific capacity of 200 mAh/g and high voltage is extensively used material for negative electrode for Li-ion battery (Tarascon & Armand, 2011, pp. 171–179). However, there are significant potential risks due to the dendrite crystal-induces short circuit rapid charging (Guo, Zhao, & Yue, 2015;

Wang et al., 2015). For the applications of grid energy storage, the initial high cost limits the extensive use of Li-ion batteries. There are certain safety risks in Li-ion batteries due to increase in temperature with internal short circuit or over-charging (Li, Yang, Zhang, & Zhang, 2015). Fig. 6.1 shows the schematic diagram of a typical lithium-ion battery.

Li-ion batteries are playing important role in development and promotion of electrical vehicles. Besides, these batteries are ideally being used in small electronics applications as a good power source as well as their applications in renewable energy grid systems and microgrid systems are being tested and demonstrated.

6.2.1.1 *Lithium-O_2 battery*

The nonaqueous $Li-O_2$ batteries possess high energy density value of ~3550 Wh/kg theoretically, which is quite higher in comparison to Li-ion batteries with density value of ~387 Wh/kg. Such high value of energy density of these batteries makes them suitable for renewable energy storage applications (Chen, Freunberger, Peng, Fontaine, & Bruce, 2013; Wu et al., 2017; Xiao et al., 2011; Yi, Liu, Qiao, He, & Zhou, 2017). The reason behind this is that anode made of lithium metal and oxygen working as cathode can freely be availed from the environment while reducing the battery's overall weight. Abundant availability of oxygen gas as cathode makes it great opportunity for its further development in future. A typical scheme of $Li-O_2$ consists of an anode of a lithium metal, cathode of porous carbon and conductive electrolyte of Li+ion. The fundamental electrochemical reaction relies on the reversible formation and decomposition of Li_2O_2 upon cycling with an equilibrium voltage of 2.96 V versus Li/Li+ and it is given as: $2Li^+ + 2e^- + O_2 \leftrightarrow Li_2O_2$ (Liu, Li et al., 2015). The schematic diagram of $Li-O_2$ battery is shown in Fig. 6.2.

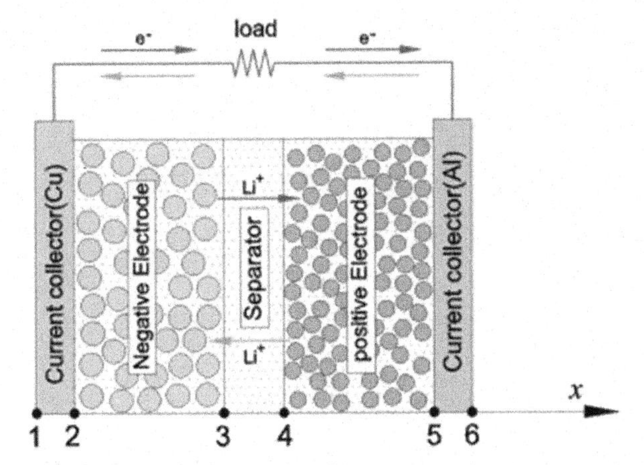

FIGURE 6.1 A schematic of typical lithium-ion battery (Yang et al., 2019).

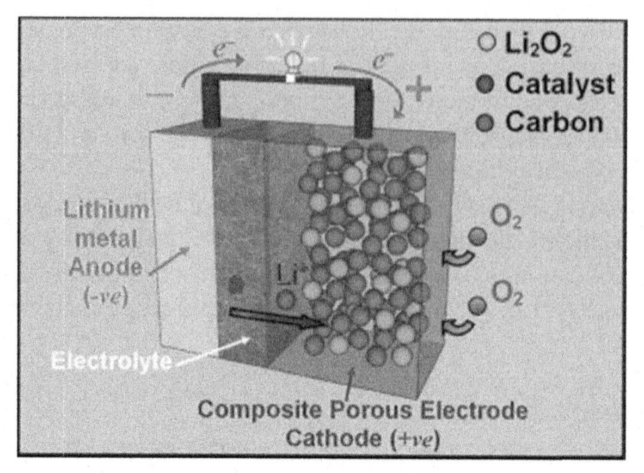

FIGURE 6.2 A diagram of type of lithium—O_2 battery (Scrosati & Garche, 2010).

6.2.1.2 Lithium cobalt oxide battery

A cobalt oxide cathode and a graphite carbon anode are used in these lithium cobalt oxide ($LiCoO_2$) batteries. Fig. 6.3 shows Schematic illustration of a lithium-ion battery. The anode (graphite) and the cathode ($LiCoO_2$) are separated by a nonaqueous liquid electrolyte (Xia, Luo, & Xie, 2012). The use these batteries in applications such as laptops, digital camera and phones is suitable choice due to their high specific energy as these can be operated between 3 and 4.2 V. These batteries can withstand charge-discharge cycles of up to 1000 before the significant drop in their performance. The energy density of battery is minimum as 150 and maximum as 200 Wh/kg. However, special cells can reach up to specific energy density of 240 Wh/kg. Thus, there are few

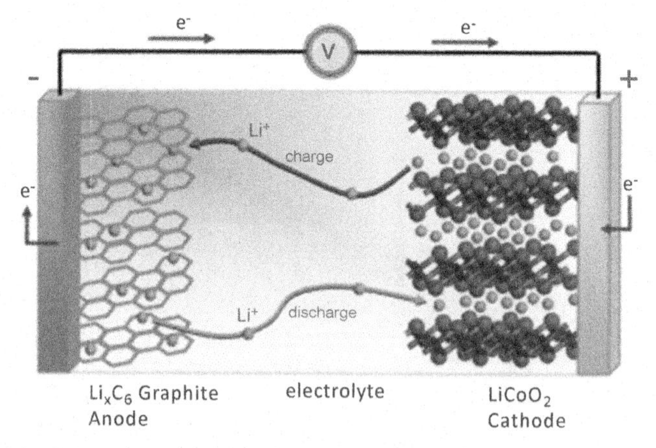

FIGURE 6.3 Schematic illustration of a lithium-ion battery. The anode (graphite) and the cathode ($LiCoO_2$) are separated by a nonaqueous liquid electrolyte (Xia et al., 2012).

main disadvantages of these batteries such as limited specific power capabilities, short life span and low thermal stability that may cause in overheating of these batteries.

6.2.1.3 *Lithium manganese oxide battery*

In these batteries, the cathode is made of lithium manganese oxide material. These are designed in a structure of 3-dimensional spinel, which causes efficient flow of ions on the electrode, which makes it high in thermal stability and safety. Therefore, these batteries are highly recommended for high load applications such as power tools and electric vehicles. The main disadvantages of these batteries are low life of 700 cycles approximately. The variation in energy density is between 100 and 150 Wh/kg. Furthermore, these batteries can be operated between 3 and 4.2 V.

6.2.1.4 *Lithium nickel manganese cobalt oxide battery*

The combination of nickel and manganese cobalt at the cathode makes these batteries very successful as these are flexible for power (lower capacity, higher current) energy (higher capacity, lower current) applications. Such advantage of flexibility makes these batteries ideal for broader applications such as usage in electric vehicles to industrial and medical devices etc. Furthermore, these are cost effective than the Li-ion batteries because of nickel at the cathode, which is cheaper in cost as compared to cobalt. These can be operated in the range of 3 and 4.2 V like Li-ion batteries. Battery life reaches up to 2000 cycles and minimum and maximum energy density is as 150 and 220 kWh/kg.

6.2.1.5 *Lithium iron phosphate battery*

In these batteries, the cathode material is phosphate, which is good for its low resistance and electrochemical performance. There are many advantages of these batteries such as good thermal stability, cycle life, lower stress at high voltage even for longer time, high tolerance at full charge conditions, high current rating and long cycle life. These advantages make them able to be used for high load currents and endurance such as batteries in vehicles in place of lead-acid batteries. However, main disadvantage is their self-discharge as compared to lithium-based batteries along with lower nominal voltage of about 3.2 V, making them low in specific energy density to about 90–120 kWh/kg. These batteries can reach up to about 2000 cycles.

6.2.1.6 *Lithium nickel cobalt aluminum oxide battery*

These batteries are similar to lithium nickel manganese cobalt oxide batteries. They can reach up to 500 cycles and can be operated in the voltage range of 3–4.2 V. These batteries possess high energy of density as 200–260 kWh/kg. However, these batteries are costlier than Li technology and their use for industrial and electric power trains is common.

6.2.1.7 Lithium titanate battery

Unlike Li-ion batteries, which use graphite material in anode the lithium titanate batteries use anode made of titanate material and material for cathode used is as in lithium nickel manganese cobalt oxide. Li–titanate is observed to perform well at high temperatures. When faced to fast charging or charging at low temperatures, there is no lithium plating or solid electrolyte interface film formation, which proves its good performance at these conditions. The applications in particular for these batteries are uninterruptible power supplies and powertrains. However, there are few main disadvantages as its low specific energy density, which is 50 Wh/kg and high cost with typical operating range of 1.8–2.85 V. They can reach cycles up to 7000.

6.2.1.8 Li–sulfur battery

As sulfur is naturally enough to resource and environmentally friendly with a high gravimetric theoretical capacity of 1672 mAh/g; therefore, Li–sulfur (Li–S) batteries are the promising batteries in performance and cost effectiveness (Xia et al., 2012; Evers et al., 2012; Zhou et al., 2015; Wu et al., 2015). By the redox reactions of S8 + 16 Li $\leftrightarrow$ $8Li_2S$ ($\sim$2.15 V vs. Li/Li+), sulfur changes its structure during charging and discharging processes (Liao et al., 2016). Li–S batteries can prove to be better option in future after further development for renewable energy storage applications. Fig. 6.4 shows the schematic diagram of a typical lithium-sulfur battery.

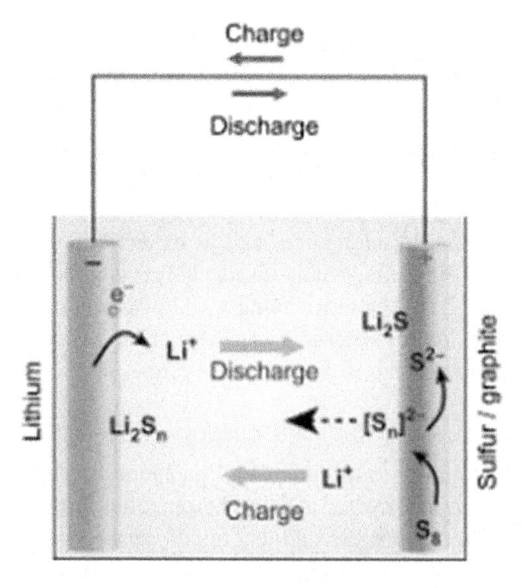

FIGURE 6.4 Lithium sulfur battery (Kurzweil & Garche, 2017).

6.2.1.9 Li–iodine battery

These batteries work as Li–redox batteries and combine the advantages of iodine-based catholyte and Li anode for achieving the energy density in higher values >250 Wh g^{-1} and high-power density along with faster charging and discharging rates. These batteries are flexible in operation and show other advantages such as the high-output voltage of ~3.05 V versus Li+/Li and higher theoretical capacity of 211 mAh/g. Such advantages of Li–I batteries makes them to be considered for applications in the large-scale energy storage and long-range electric vehicles (Chen & Lu, 2016; Liu, Shadike, Ding, Sang, & Fu, 2015; Zhao & Byon, 2013; Zhao et al., 2014; Zhao, Wang, & Byon, 2013). Fig. 6.5 shows the schematic cut-view of a lithium-iodine battery.

6.2.2 Flow batteries

These batteries are one of the electrochemical cell types, which are combination of fuel cell and conventional battery, wherein energy is provided with circulation of two electrolytes (liquids with metallic salts). These electrolytes are circulated through the core consisting of negative and positive electrode, which are separated by membrane. The exchange of ions between the catholyte and anolyte is produces with the circulation of electrolytes resulting in generation of electricity and reversing the same process causes the charge in battery. Unlike the conventional batteries which store the energy in electrodes the flow batteries stores energy in the electrolyte. Capacity of the battery can be determined by the volume of electrolyte.

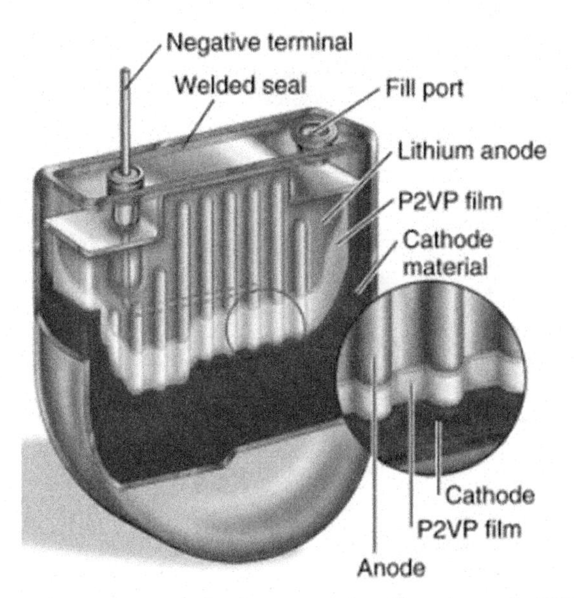

FIGURE 6.5 Lithium-iodine battery cut view (Untereker et al., 2017).

6.2.2.1 *Redox flow battery*

The most common type of the flow batteries are reduction-oxidation batteries. These batteries generate the electricity through the potential difference of the tanks. The solution contained by the both tanks becomes same containing both positive and negative ions. Vanadium–polyhalide, vanadium–vanadium, hydrogen–bromine, bromine–polysulfide, and iron–chromium are the common materials being used in these batteries. There are one or more electroactive components used as solid layer in hybrid flow batteries. On fuel cell electrode and one battery electrode are used in electrochemical cell and Lead-Acid, Zinc–Bromine and Zinc–Cerium are the materials used in these batteries. Fig. 6.6 shows the schematic diagram of redox flow battery.

6.2.2.2 *Vanadium-based flow battery*

In these batteries, sulfuric acid is used as electrolyte, which stores electrical energy in the chemical energy from. There are various states of vanadium ions in sulfuric acid electrolyte and electrons are transferred during the operation of the battery through the change of valence state of the vanadium. In this way, the conversion of electric and chemical energy is completed (Li & Li, 2011). Currently, the critical materials of the battery such as ion exchange membranes, electrolytes and electrodes are the focus for research in these batteries (Pour et al., 2015; Schweiss, 2015; Yang, Ye, Cheng, & Zhao, 2015). These batteries have many advantages such as high power, long life, high efficiency, high safety and high capacity. Because of these advantages, these batteries are rapidly developed in the short time. However, there are still many issues for its commercialization/industrialization due to the critical materials e.g., electrodes,

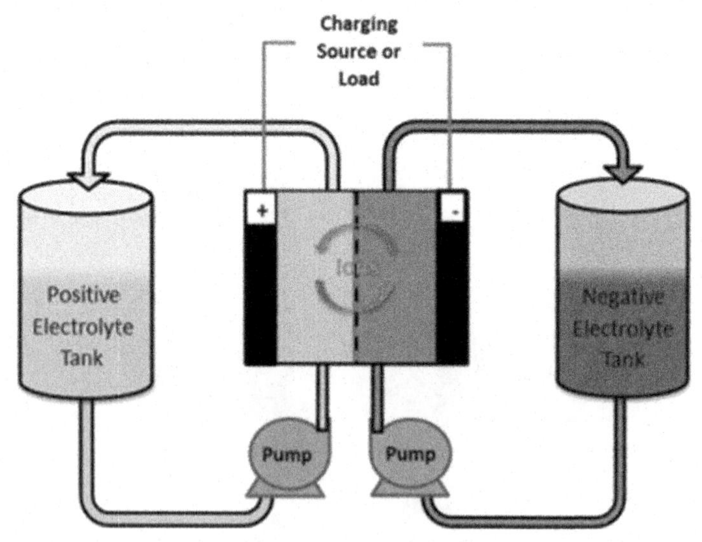

FIGURE 6.6 Schematic of redox flow battery (Bamgbopa, Almheiri, & Sun, 2017).

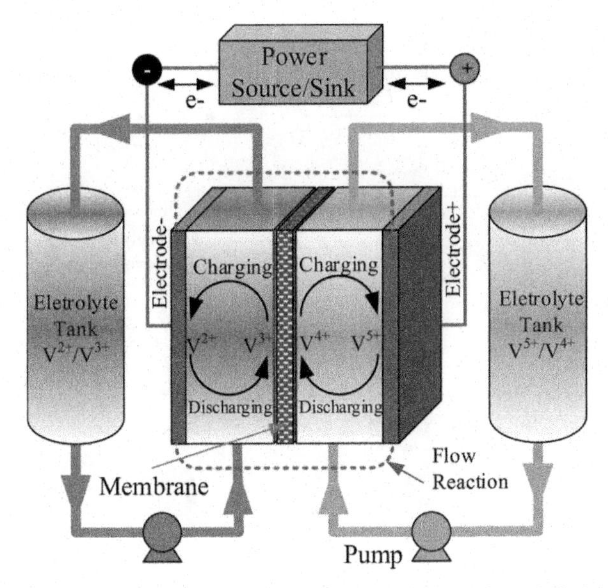

FIGURE 6.7 Vanadium redox flow battery (Hannan, Hoque, Mohamed, & Ayob, 2017).

electrolytes and ion-exchange membranes along with their high capital for construction (Mohamed, Leung, & Sulaiman, 2015). Fig. 6.7 shows the schematic diagram of a vanadium-based flow battery.

6.2.2.3 Membrane-less flow battery

These batteries don't use the membranes which are used for the purpose of separating two electrolytes from each other. Therefore, such purpose of separation of the electrolytes is served by maintaining the laminar flow. There are different power densities depending on the systems such as 800 W/m^3 for vanadium-vanadium to >1000 W/m^3 for zinc−bromine and lead-acid systems. Further, for hydrogen-lithium systems, it can reach up to 15,000 W/m^3. Depending on materials and technology, there are different voltages for these battery cells, which range between 0.5 and 2.4 V. There is another advantage of these batteries, which is that power and energy requirements can be manipulated because the electrodes are not the part of electrochemical fuel. Furthermore, these batteries are stable in performance with longer lifetime and more durable as the electrodes do not possess active material and the separation of these active materials provides safer whole system. These flow batteries are also having the advantage of deep discharging, which doesn't affect the cycle life. This makes them to reach unlimited cycles of charging without any damage or impact on their capacity. However, size of the batteries is the main disadvantage even thought their construction is so simple. Due to the size, it limits the applications of these batteries for industrial applications, pump systems, vessels and sensor etc. Fig. 6.8 shows the schematic diagram of membrane less flow battery.

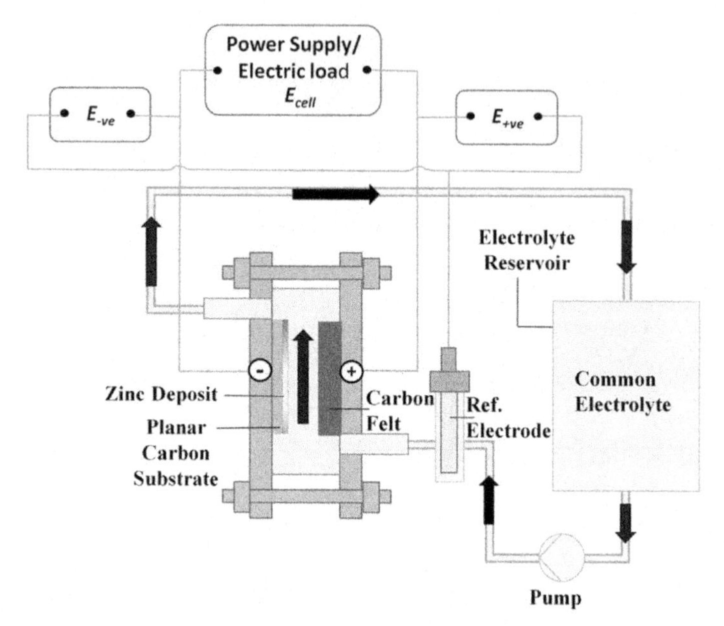

FIGURE 6.8 A type of membrane less flow battery (Leung, Martin, Shah et al., 2017).

6.2.2.4 Dual-liquid redox battery

Dual-liquid redox batteries store energy though two liquid redox-couple electrolytes; electrolyte and catholyte for to avoid the metal anode safety issues where, dual electrolytes can be in external reservoir at different volumes. By using the pumps for mixing electrodes and reservoirs, these batteries can create high-rate diffusion (Kim et al., 2015; Wang, He, & Zhou, 2012). The batteries are flexibly operated with moderate cost and are transportable amongst the technologies used for applications of storing energy at grid scale because of the modular design of the batteries. Thus, these batteries with acolyte and catholyte can be used for different applications of varying requirements with huge flexibility (Li, Weng, Zou, Cong, & Lu, 2016). Fig. 6.9 shows the schematic diagram of dual-flow battery.

6.2.3 Nickel-based batteries

Nickel-based batteries were invented in the 19th century and since then many advancements are carried out to improve this technology. Porous nickel electrode is used in these for the deposit of active materials. Types of the Ni-based batteries are given below. Fig. 6.10 shows the schematic of Nickel-based battery using cadmium.

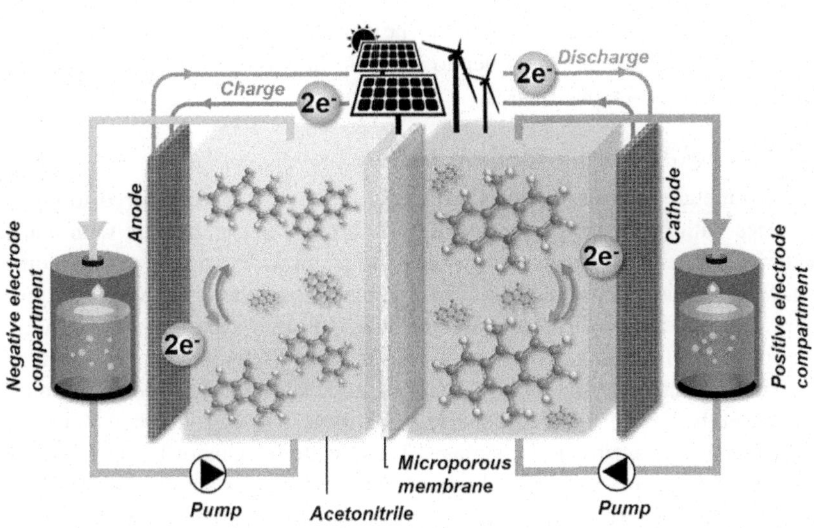

FIGURE 6.9 Schematic of dual-flow battery and cell configuration (Kwon et al., 2018).

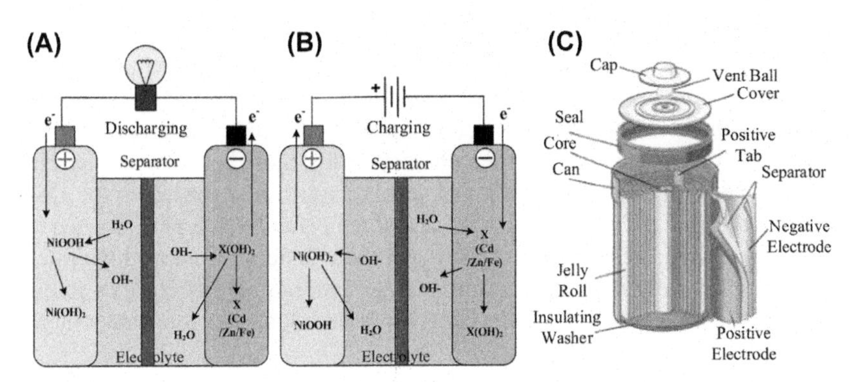

FIGURE 6.10 Nickel-based battery (Hannan et al., 2017).

6.2.3.1 Nickel–cadmium battery

A nickel–cadmium (NiCd) battery has good performance at low temperatures, is cost-effective, and supports fast charging. Furthermore, if properly maintained, it can reach more than 1000 cycles, showing its high durability. Such properties of the battery have led it to be right choice for use in wind energy systems stabilization and aviation industry for over many years. However, very big drawback of these batteries is toxic material of cadmium, which is not safe. Cadmium can be not disposed of in the lands due to its toxicity, which may cause damage to soil and create pollution. Therefore, other safer technologies than this cadmium-based battery have been utilized. These batteries have memory effect, which causes them to be fully charged and discharged after some periods repeatedly in order to address this issue. The specific energy density is low at 45–80 Wh/kg.

6.2.3.2 Nickel–metal–hydride battery

Nickel–metal–hydride batteries are newer technology as compared to NiCd batteries with high energy density of ~40% more in comparison with NiCd batteries. Better performance in easy recycling and wide range of temperatures makes them more advantageous. The applications of these batteries are typically in preparation of rechargeable batteries of sizes such as AA and AAA etc. for electronics usage purposes. However, high discharge rate e.g., it drains 20% of its energy in first 24 h of charging and 10% drainage for each month subsequently, is the main disadvantage of these batteries making them not suitable for long-term energy storage. These batteries cannot be overcharged as these are sensitive for overcharging, which demands smart algorithm for charging to keep them safe. These can reach up to 500 cycles and have typically 60 to 120 Wh/kg of specific energy density.

6.2.3.3 Nickel–iron battery

Nickel–iron batteries are resilient to overcharging and discharging along with high temperature and vibrations resistance. In these batteries, the electrolyte is made of potassium hydroxide, anode is made of iron and cathode is made of oxide-hydroxide. According to their characteristics, these batteries are used in railroad signaling, trucks/forklifts and mines. However, there are some disadvantages in these batteries such as at low temperatures the performance is lower, low energy density of 50 Wh/kg and these battery discharge high rates of about 40% every month. In addition, these batteries are not cheaper and if compared to lead acid and Li-ion batteries these are almost four times more expensive.

6.2.3.4 Nickel–zinc battery

Nickel–zinc batteries are comparatively better than NiCd batteries and both batteries use alkaline electrolyte. Nickel–zinc batteries have better voltage of 1.65 V in comparison with NiCd batteries, which have voltage of 1.2 V. Furthermore, these batteries are not toxic like NiCd, which makes them comparatively better than NiCd. These batteries can go up to 300 cycles and have specific energy density of about 100 Wh/kg.

6.2.3.5 Nickel–hydrogen battery

The development of the nickel-hydrogen battery was carried out due to the metal instabilities issues in nickel–metal–hydride batteries. Solid nickel and hydrogen are used as electrodes. The electrodes, electrolyte and screen are confined steel canister at high pressure of 8270 kPa. The advantages of these batteries are low self-discharge and can work with better performance in great range of temperatures e.g.,−28 to +54°C along with longer service life. Due to such advantages, the main applications of these batteries are in satellites. There is 40−75 Wh/kg specific energy density in these batteries.

6.2.4 Metal-air batteries

Metal-air batteries structure is comprised of an anode of pure metal, cathode of air, electrolyte and a separator, which is an insulator and allows the transfer of ions only. It functions in a way that oxidation reactions occur during discharge to the metal and it dissolves in the liquid electrolyte. Further, induction of oxygen reduction occurs in the air cathode. These batteries possess much higher specific energy capacity, which may reach up to 12,000 Wh/kg approximately, comparable to the energy of petrol. This high energy is because of battery configurations open to air, which plays role as the reactant in the battery. Thus, it makes this battery competitive candidate for automotive industry. However, due to some challenges, these batteries have not been used for commercial purposes. There are several types of metal-air battery technologies such as Li−air, Mg-air, Na-air, zinc−air, K-air, iron-air, Al-air, and sodium-air. Li−air is the most promising, among others. Some batteries types are given below. Fig. 6.11 shows the diagram of metal-air battery.

6.2.4.1 Zinc−air battery

Zinc−air batteries have attractive theoretical energy density of 1086 Wh/kg including oxygen but it is quite lesser than Li−air batteries, which is 1910 Wh/kg (Imanishi & Yamamoto, 2014). These batteries can be manufactured at low costs due to zinc compatibility with an aqueous alkaline as compared nonaqueous-based cells. However, development of zinc−air batteries is slowed down due to zinc tendency of forming dendrites after charge-discharge cycle repetitions along with low power outputs (Li & Dai, 2014). In addition, carbon

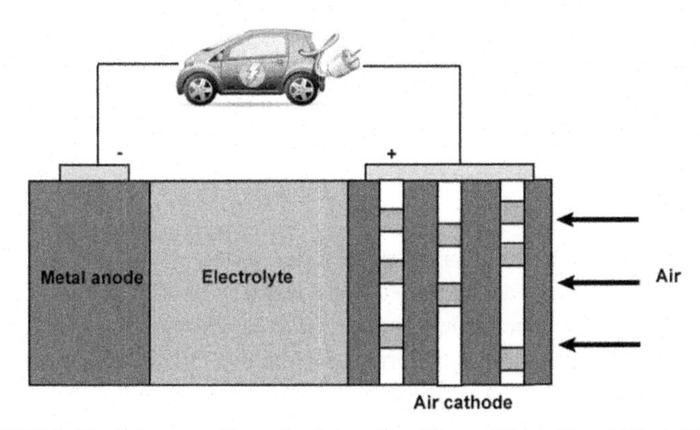

FIGURE 6.11 Schematic of metal−air batteries (Zhang, Wang, Xie, & Zhou, 2016).

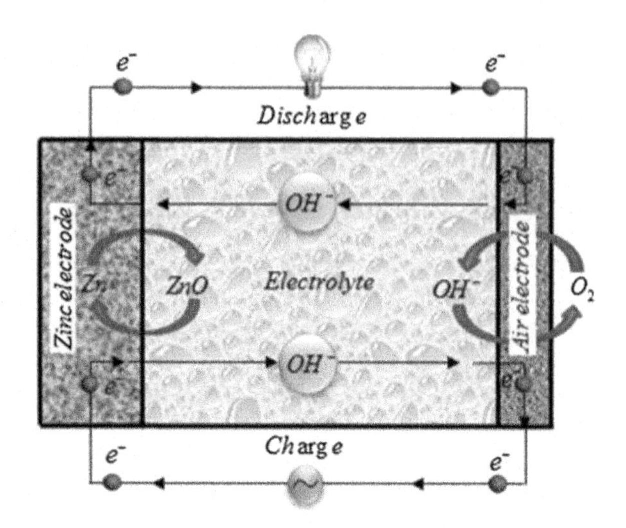

FIGURE 6.12 Schematic diagram of zinc–air battery (Pei, Wang, & Ma, 2014).

dioxide can absorb into solution, which makes it insoluble compounds which block electrode making electrolyte low in conductivity and low battery performance. Therefore, air purification system should be considered while designing such batteries. Schematic diagram of a typical zinc–air battery is shown in Fig. 6.12. The technology of these batteries is mature enough and is being used in medical and telecommunications applications commercially.

6.2.4.2 *Iron–air battery*

Iron–air batteries also work with an aqueous alkaline electrolyte. These batteries are better in terms of tripping or redisposition issues; however, these batteries have lower energy density theoretically, which is 764 Wh/kg as compared to zinc–air batteries. If compared with Pb-acid or NiFe batteries, iron–air batteries are higher energy density. Furthermore, rechargeable iron–air batteries show relatively low energy efficiencies (ca. 35%). Similar to zinc–air batteries, iron–air batteries need the further development for efficient oxygen electrodes (McKerracher et al., 2015). Fig. 6.13 shows the schematic diagram of a typical iron–air battery.

6.2.5 Lead-acid batteries

Lead acid batteries are secondary type batteries, wherein sulfuric acid function as electrolyte and the lead and its oxides function as electrode (Wei et al., 2016). For the reliable operation of power systems, Pb-acid batteries have been used in the power plants and transformer substations as backup for many years (Cheng et al., 2007). However, these batteries have not been recently used for energy storage even though these are cheaper due to environmental pollution.

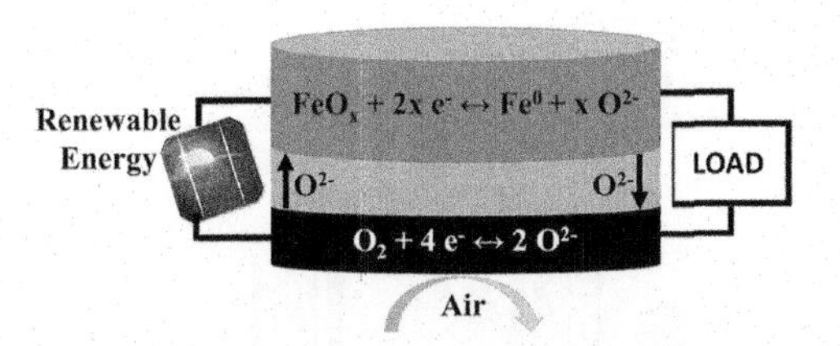

FIGURE 6.13 Functioning of Fe-air battery (Trocino, Lo Faro, Zignani, Antonucci, & Aricò, 2019).

These batteries other disadvantages of low power density, low cycle life, low energy density, low charging times and high self-discharge rates etc. Furthermore, if deeply or rapidly discharged, their energy capacity is affected greatly. For the cycle life, it can be increased by adding activated carbon to negative plate of lead acid battery (Shukla, Venugopalan, & Hariprakash, 2001). The battery is classified in to two types due to different methods of activated carbon addition into electrodes. One configuration is called ultra-battery and the other as an advanced lead acid battery (Hang et al., 2005; Kao & Chou, 2010; Kolthoff & Tomsicek, 1935; Leung, Li, Liras et al., 2017; Li et al., 2016). The lead-carbon battery is has more specific power and enhanced cycle life at low discharge depths but there are no difference in the terms of specific energy between the lead-carbon and acid-battery (Zhang, Wei, Cao, & Lin, 2018). Fig. 6.14 shows the schematic diagram of a typical lead-acid battery.

6.2.6 Sodium–sulfur batteries

Sodium–sulfur batteries have the basis of molten salt technology, where, molten sodium and molten sulfur are used as negative and positive electrodes, and solid ceramic sodium alumina acting as electrolyte separates these two electrodes in these batteries (Dunn, Kamath, & Tarascon, 2011). Electricity is generated in a way that during discharge, sodium metal atoms release the

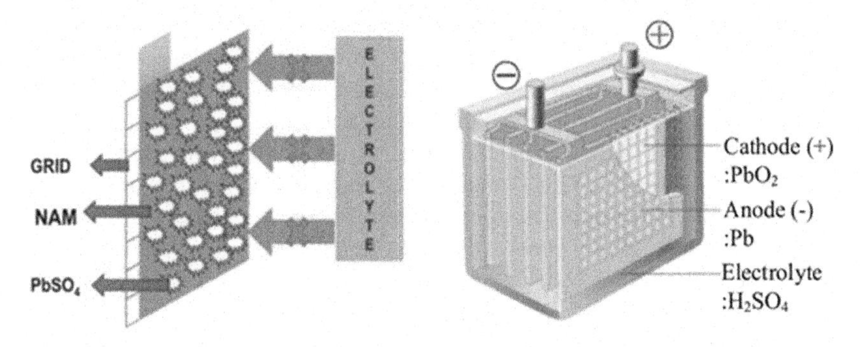

FIGURE 6.14 Lead acid battery (Vangapally, Jindal, Gaffoor, & Martha, 2018).

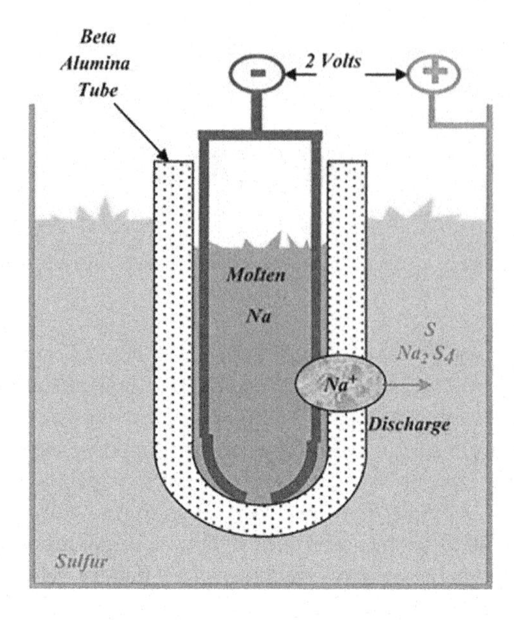

FIGURE 6.15 Schematic of a Na–S battery (Kumar, Rajouria, Kuhar, & Kanchan, 2017).

electrons, which form the sodium ions moving to the positive electrode through the electrolyte. The specific energy density is about 150 Wh/kg or some times more higher than this value and high round-trip efficiency of ~90% (Manthiram & Yu, 2015). The applications of these batteries are mostly peak shaving, renewable energy stabilization and provision of services of secondary importance (Xin, Yin, Guo, & Wan, 2014; Yu & Manthiram, 2015). Fig. 6.15 shows the schematic diagram of a sodium–sulfur battery.

6.2.7 Aluminum-ion batteries

Aluminum-ion batteries function as the electrochemical disposition and dissolution of aluminum at anode, and the intercalation/de-intercalation of chloraluminite anions in the graphite cathode. Practically, these batteries have the power density of 3000 W/kg and energy density of 40 Wh/kg making them to be similar to lead-acid batteries in such characteristics. There is little drop in capacity of the batteries after the stable charge and discharge cycles of 7500 (Angell et al., 2017; Jiao, Wang, Tu, Tian, & Jiao, 2017; Jung, Kang, Yoo, Choi, & Han, 2016; Lin, Gong et al., 2015; Wu, Xu, Chen, & Ouyang, 2016; Yu et al., 2017). Further advantages of these batteries are safety, cost effectiveness and high power density. Furthermore, these batteries are under research for ultrafast charge and discharge along with grid energy storage applications (Prentis, 2016; Ye, 2014). Fig. 6.16 shows the schematic diagram of an aluminum–iron battery.

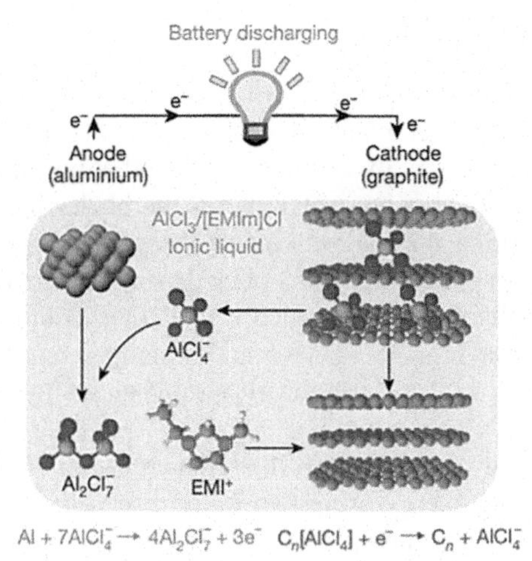

FIGURE 6.16 Schematic of aluminum-ion battery (Zhang et al., 2018).

6.2.8 Copper–zinc batteries

Rechargeable copper–zinc battery is another important technology, which uses aqueous electrolyte by "Cumulus Energy Storage." This technology is based on processes used in metal refining. This project of developing copper–zinc batteries is focused on cost effectiveness and safety the systems. Furthermore, the capacity design for these batteries is aimed at a range from 1 to 100 MWh. Fig. 6.17 shows the schematic diagram of copper–zinc battery.

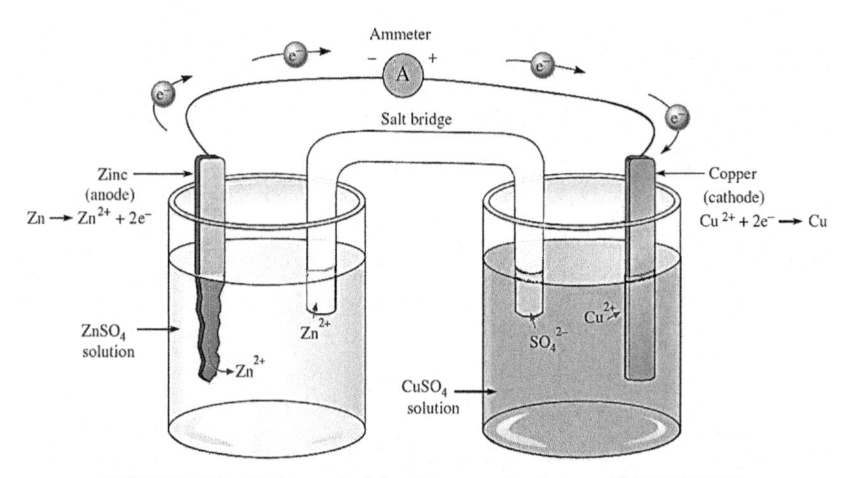

FIGURE 6.17 Working principle of copper–zinc battery (XAKTY, 2019).

6.3 Hydro energy storage

Hydro energy uses the kinetic energy of flowing and falling water to create electricity in a hydropower plant. Hydropower plants use the turbines and generators to produce electricity as similar to the other power-generating units. The kinetic force of the moving water causes the blades of a turbine to spin, which results in turning the magnets inside the generator and produce electricity. The generation of electricity of a plant depends on the volume of flowing water and the available head from which it falls. One example of such hydropower plant is pumped storage where water is pumped into the impoundment during off-peak times so that it can be released when the hydropower power is needed.

Pump hydro energy storage (PHES), as shown in Fig. 6.18, is a process of storing and generating energy using two water reservoirs or basins located at different elevations by pumping water from a lower reservoir to a higher reservoir (Poulain, de Dreuzy, & Goderniaux, 2018). In a PHES plant, when electricity is relatively cheap (i.e., electricity during off-peak hours), the operating of the pump is carried out using the electricity generated from other energy sources to raise the water from the lower reservoir to the upper one. When the demand for electricity is high, the water from the upper reservoir flows back into the lower reservoir through the turbine to generate electricity. The assemblies of a reversible turbine/generator can act as a pump or a turbine as required. PHES systems coupled with wind and solar photovoltaic power are shown in Figs. 6.19 and 6.20, respectively (Rehman, Al-Hadhrami, & Alam, 2015). A hybrid solar–wind system with PHES is also illustrated in Fig. 6.21 (Ma, Yang, Lu, & Peng, 2014).

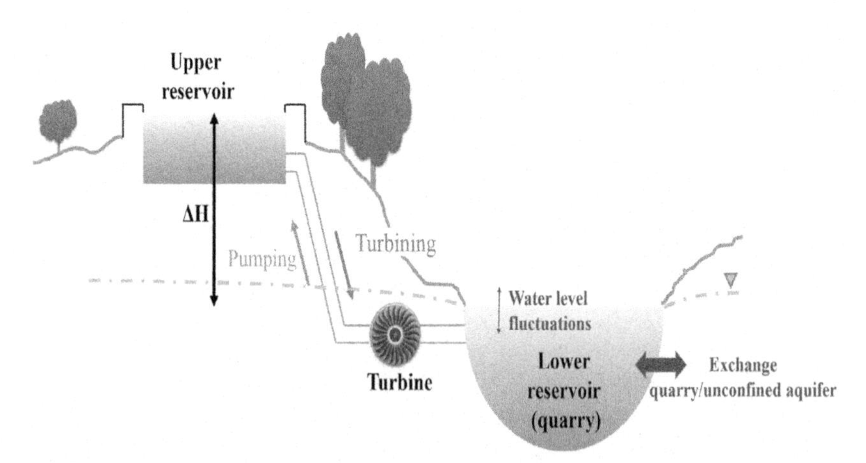

FIGURE 6.18 Pump storage system between an upper artificial reservoir and a lower reservoir (Poulain et al., 2018).

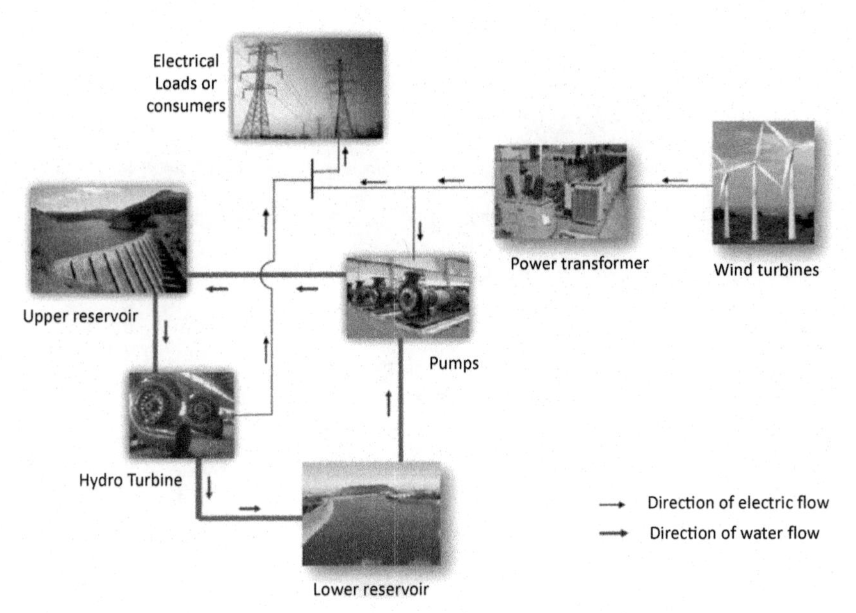

FIGURE 6.19 Wind power-based pumped hydro energy storage system (Rehman et al., 2015).

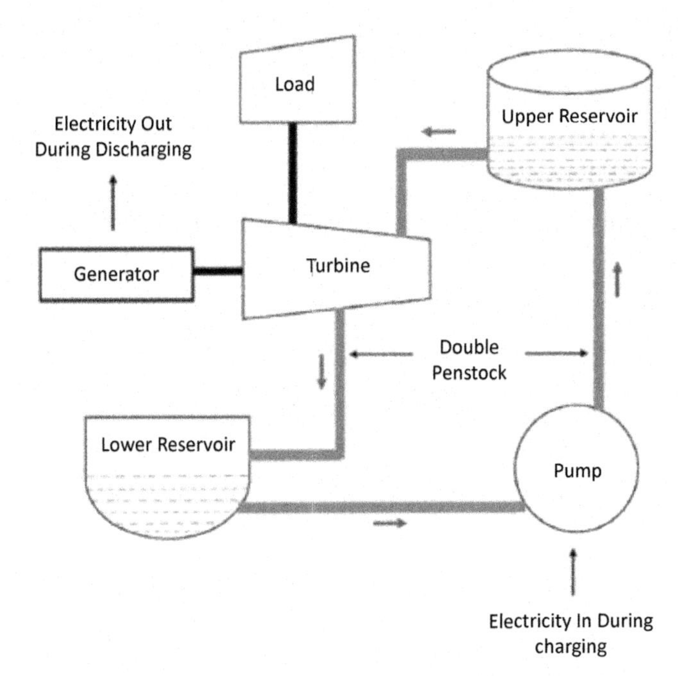

FIGURE 6.20 Solar photovoltaic power-based pump hydro energy storage system (Rehman et al., 2015).

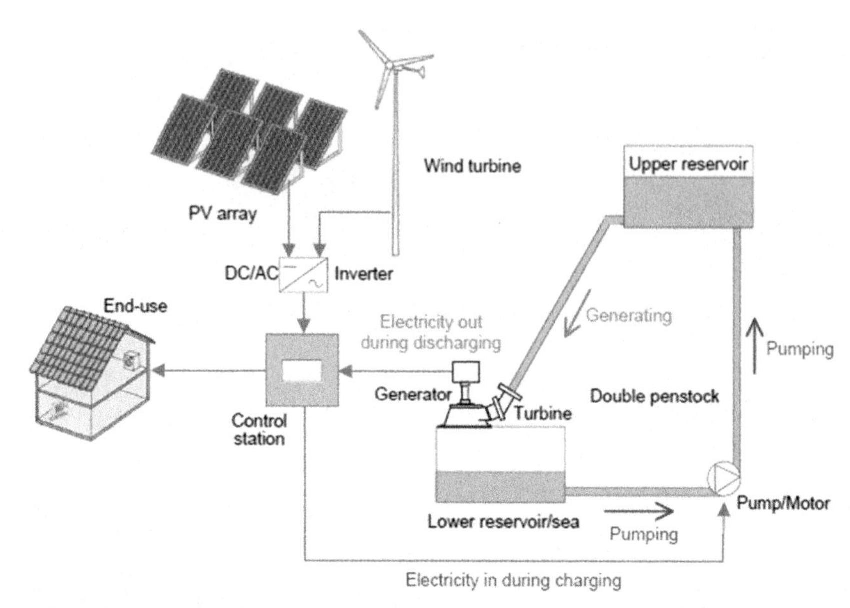

FIGURE 6.21 A hybrid solar–wind system with pumped storage system (Ma et al., 2014).

The PHES is the most mature large-scale energy storage technology available across the world. It provides the electrical storage capacity of about 99%, with a total installed capacity of more than 120 GW (Zafirakis, Chalvatzis, Baiocchi, & Daskalakis, 2013) and contributes to 3% power generation (Energy storage – packing some power, 2012). The PHES offers a very low energy density, almost zero self-discharge, reasonable price per stored energy unit, a high round-trip efficiency, a few seconds response time and very fast start-up time (Luo et al., 2015; Shyam & Kanakasabapathy, 2018). However, the major disadvantages of the PHES plant are associated with the need for sufficient acceptable water availability and appropriate geographical morphology. So in the countries where favorable morphology is available, new PHES units can be installed. It is also possible to upgrade conventional hydroelectric power plants in conjunction with the PHES (Benato & Stoppato, 2018).

The basic components of the PHES system are

i. upper reservoir or forebay
ii. penstock
iii. motor generator
iv. pump turbine
v. tail race
vi. lower reservoir

6.3.1 General classification of pump hydro energy storage plant

According to the water regime, PHES can be classified as:

(i) Closed-loop: This type consists of upper and lower reservoirs, and none of the reservoirs is connected to another source of water.
(ii) Semi-open: This type is made of one lake or river or sea as a reservoir and another artificial or modified reservoir with a continuous through flow.
(iii) Open-system (pump-back): This type of plant undergoes the flow of water continuously through both the upper and lower basins (Wänn et al., 2012).

According to the power capacity of PHES, this system can be classified into micro, mini, small, and large plants.

(i) Micro: This plant has the capacity in the range of 1−100 kW.
(ii) Mini: The PHES systems having installed capacity from 100 kW to 1 MW are generally known as the mini plant.
(iii) Small: Typical capacities of these plants are between 1 MW and 10−30 MW.
(iv) Large: Large PHES plants have a capacity of more than 10 MW or 30 MW. Note that the upper limit may differ from country to country (Breeze, 2018a).

6.3.2 Performance analysis of a pump hydro energy storage plant

The efficiency of pump hydro energy storage can vary depending on the various components of the plants such as waterways and electrical-mechanical components. The performance analysis of a typical PHES plant can be done following the sections below (Canales & Beluco, 2014; Koko, Kusakana, & Vermaak, 2018; Ma et al., 2014).

The PHES system consists of a separated motor/pump unit and a turbine/generator unit. The pump stores a certain amount of water from the lower reservoir into the upper one.

The power output (P) that can be derived from a given quantity of water is directly proportional to the head (H) and the flow rate (Q). For elevating the water into the upper reservoir, the supplied power to the motor/pump unit is

$$P_{mp} = \frac{\rho_w g H Q_{mp}}{\eta_{mp}}. \tag{6.1}$$

where ρ_w is the density of the water (1000 kg/m^3), g is the gravitational acceleration (9.81 m/s^2), H is the height of the water head (m), Q_{mp} is the volumetric flow rate of the water (m^3/s), and η_{mp} is the efficiency of the motor/pump unit.

When there is a shortage in power supply during high demand for it, the available water can be released from the upper reservoir for operating the turbine for electricity generation. The power produced by the turbine/generator unit is defined by

$$P_{tg} = \rho_w g H Q_{tg} \eta_{tg} \tag{6.2}$$

where Q_{tg} is the flow rate of water through the turbine (m^3/s) and η_{tg} is the efficiency of the turbine/generator unit.

The energy output (E) of the PHES plant is proportional to the storage capacity (V) of the water available in the upper reservoir. In general, 1 m^3 of water falling 100 m can generate 0.272 kWh of electricity (Ibrahim, Ilinca, & Perron, 2008; Kapila, Oni, & Kumar, 2017). The energy output (E) of the PHES plant is estimated by

$$E = \frac{V \rho_w g H \eta_{tg}}{3.6 \times 10^6} \tag{6.3}$$

where V is the storage capacity of the upper reservoir (m^3).

The total efficiency of a pumping system (η) can be estimated considering efficiencies like the efficiency of pump (η_p), the efficiency of transmission between the prime mover and the rotating equipment (η_t), and the efficiency of motor (η_m). This can be defined as (Al Zohbi, Hendrick, Renier, & Bouillard, 2016)

$$\eta = \eta_p \eta_t \eta_m \tag{6.4}$$

Due to the system losses (i.e., pipe friction losses, losses associated with turbine and pump, and so on), the PHES plants are generally able to recover approximately 70%–80% of power inputs (Yang & Jackson, 2011). The majority of losses are from pump/turbine units in both pumping and generating modes.

6.3.3 Design considerations for pump hydro energy storage plants

Numerous components required to be considered during the design of a new PHES plant. However, the preliminary considerations include head, flow rates, waterways, two reservoirs, pump/turbine selection, plant capacity, and so on.

6.3.3.1 Head

Existing hydro energy storage plants have height in the range of 30–750 m. The greater the head, the more energy water can store. The high head plants are more cost-effective than the low head plant due to the need for smaller equipment. The plant with high head often requires more complex multistage pump/turbine units to elevate water to the upper reservoir (Breeze, 2018b). The high head plants also require smaller reservoirs and pumps/turbines units than low head to produce an equal amount of power, so the high heads are preferred.

6.3.3.2 Flow rates

The available head and flow rates are estimated based on the desired generating capacity of the plant. Larger conduits and pump/turbine units are required for the high flow rate of water at a particular head. It is necessary to ensure the minimum head losses in the conduits or waterways. In this regard, the conduits with large diameter can be used rather than the smaller diameter conduits. However, larger conduits are typically more expensive to construct because they require more civil works.

6.3.3.3 Waterways

The design of the waterways affects the overall efficiency of the PHES plant and the performance of the pump/turbine units. The waterways are a system of pipes that connect the upper reservoir, pump/turbine units, and lower reservoir. The waterways consist of two sections: the first section is between the upper reservoir and the pump/turbine unit(s), and the second section is between the pump/turbine unit(s) and the lower reservoir. The shortest waterway between pump/turbine units and reservoirs is preferred to reduce the construction cost and friction loss in the system. However, making this may become complex because of the large variations in the topography of the land.

6.3.3.4 Upper and lower reservoirs

The PHES plants need two reservoirs (i.e., an upper and a lower one) located at different heights. A natural source (e.g., river, lake, pond, and sea) can act as one reservoir while the other one is manufactured. Two existing lakes or two manufactured reservoirs can also be used to create a pumped storage facility. However, using two manufactured reservoirs can increase significantly the capital cost of the plant. The size of the reservoir is determined by the site's topographic and hydrological conditions. It is necessary to ensure at the design stage whether one or both reservoir sizes are to be made. The storage capacity of the PHES plant is determined based on the total quantity of water both the reservoirs hold. The more hours of power can be stored if the reservoir size is larger for a given turbine. Some plants may have a storage capacity that can sustain up to 20 h while others up to 4 h. The energy storage efficiency is 70%–80% when a typical PHES plant operates on a daily cycle (Breeze, 2018b).

6.3.3.5 Turbine selection

A PHES plant must be able to produce both power from water running into the lower reservoir and store energy by pumping water into the upper reservoir. Both pumps and turbines are required for this purpose. Either Pelton or Francis turbines can be applied, but Pelton is chosen for a plant with a high head. PHES plants can respond quickly to a demand for power regardless of the type of turbine. For instance, the Dinorwig plant in Wales has six 300 MW pump turbines which can synchronize from zero to the full power output of 1320 MW in 75 s. Even the plant with a capacity of 800 MW can be operational in 16 s when the turbines are synchronized quickly to deliver the power (Breeze, 2018b).

6.3.3.6 Plant capability

The overall capability of a PHES plant for storing and generating energy is determined by the reservoirs size and the head height between the two reservoirs. The more a given quantity of water, the more storing capability it has. The power output of PHES plants depends on the pressure imposed by water, the flow rate of the water through the turbine, and the power of the pump/motor and turbine/generator units (Luo et al., 2015). The plant with a high head can offer more power than a small head from an equivalent flow of water.

6.3.3.7 Investment costs

The investment costs include the cost associated with the land acquisition, civil works, steel structure, mechanical and electrical machines. The cost differs with the timeframe of installation, the location and size of the plant and many other factors. Grid fees for pumping electricity, water usage fees, and operational restrictions and mitigation have also to be considered in the investment cost and may affect the profitability. Furthermore, the maintenance and replacement cost may increase due to the low life span and cycling times. The typical investment cost is $1170/KW (Luo et al., 2015).

6.3.3.8 Profitability

The PHES plants can get the return of its investment by selling electricity and payments obtained from the grid services. There are three probable revenue sources to meet the investors' profitability as follows (Steffen, 2012).

6.3.3.8.1 Price arbitrage

Balancing the increasingly intermittent electricity generation with the variable demand of a load per day is the main contribution of the PHES plant. The associated income is the arbitrage spread over time, which relies on the frequency and variation in price in the peak and low demand of the load.

6.3.3.8.2 Grid services

Reserve power and grid services such as reactive power and black start ability can generate additional income for the PHES systems in addition to the price arbitrage. The reserve power can solely produce more profit than the time spread arbitrage, with an internal rate of return of 7.3% for the investment cost of $1115/kW (Steffen, 2012). It is also possible to use the reserve power along with the time spread arbitrage.

6.3.3.8.3 Capacity subsidies

There are significant uncertainties about the price of electricity, income from reserve power and other services. These uncertainties are not uncommon in the case of PHES plant, so the profitability of PHES may eventually depend on the subsidies by the government or any other organization.

6.3.4 Applications of pump hydro energy storage plant

The PHES system is a well-proven and widely adopted technology for large-scale energy storage. Wind and solar energies can be used in combination with pump storage schemes for generating electricity. There are many applications of PHES plants, which are mentioned as follows (Luo et al., 2015; Rohit, Devi, & Rangnekar, 2017):

6.3.4.1 Energy arbitrage

The generation of energy is very costly, and the PHES system can store the energy, increasing the efficiency of the system and optimizing it economically. When the price of the generated energy is less, the storage of energy can be done. The selling of energy occurs during increased demand when the price of electricity is high. In the case of renewable energy systems such as wind, solar, and wave, PHES can also store energy during off-peak hours and render the stored energy into the system during peak hours (Bragard, Soltau, Thomas, & De Doncker, 2010).

6.3.4.2 Power rating

PHES system can be applied for the management of energy with different power ratings (i.e., large-scale (above 100 MW) and medium/small scale (1–100 MW)). Typical batteries, fuel cells, and solar fuels are fit for the management of small/medium-scale energy, whereas PHES systems, due to almost zero self-discharge, are more suitable for the management of large-scale energy.

6.3.4.3 Emergency and telecommunication backup power

For an emergency power supply, PHES system can be used in the case of power failure. For example, if the power cut happens in the telecommunication sector, the PHES system can then provide sufficient power to keep it running until the main supply reinstated or enable orderly shutdown of the system. In addition, this system can be applied to the data center of any company to provide emergency power.

6.3.4.4 Time shifting

When electrical energy is less expensive, time-shifting can be obtained by storing electrical energy. After this, the stored energy can be used or sold during the high demand periods of energy. Storing energy using PHES can also facilitate the shifting of renewable energy from one-time frame of the day to another and from weekdays to weekends (Zabalawi, Mandic, & Nasiri, 2008). In this regard, PHES technology needs to have power between 1 and 100 MW. Other energy storage systems like conventional batteries and solar fuels have also the capability for such application.

6.3.4.5 Peak shaving

The peak load is covered by installing the peak shaving (Makansi & Abboud, 2002). In general, installation of it is performed at the consumer end, while arbitrage is placed at the supply end (Eyer & Corey, 2010). The economic benefits can be gained by reducing the usage of expensive peak electricity using the PHES system (Luo et al., 2015). The application of PHES also helps to improve the design of the system by supporting peak demand.

6.3.4.6 Seasonal energy storage

For the large fluctuation in energy production and consumption, and storage at different times of a day and year, technologies with large energy storage capacity and no self-discharge are essential. However, currently, no energy storage technology is available for this application. PHES and other energy storage devices like hydrogen fuel cells and solar fuels have the prospect in applications to such area.

6.3.4.7 Black start

Any unexpected events may result in power stoppage of a system or in a part, and the stability of the system can be compromised. The system can then be reinstated through a system named black start. The functions of a black start are the management of power, voltage control, and balancing the load. The PHES system can start the system from shutdown without having power from the grid and energize distribution lines.

6.3.4.8 Frequency regulation

The regulation of frequency is very important in power system for managing lots of small variations occurred in the frequency of the system. The PHES in a frequency regulator assists the power systems through correcting the fluctuations of the frequency within the allowable limits.

6.3.4.9 Standby reserve

PHES plants can act as temporary additional producing units for the middle/large-scale grid to balance supply and consumption of electrical energy. This reserve can be used for solving the situation when actual demand goes beyond the estimated demand and/or plant breakdowns.

6.3.5 Site selection for pump hydro energy storage plant

There are three main groups of factors that should be considered when selecting a site for a PHSE plant (Nzotcha, Kenfack, & Blanche Manjia, 2019):

- **Techno-economic factors**
  - power grid proximity
  - road network
  - site geologic condition
  - head-distance ration
  - gross head
  - site topology
  - earthquake proximity
- **Social factors**
  - urban area proximity
  - people settlement
  - land use
  - latest fault potential
- **Environmental factors**
  - water use
  - sealed ground surface
  - land cover
  - local wind speed
  - local solar irradiation

6.4 Thermal energy storage

Primary energy demand is mainly supplied by fossil fuels which have led this source to the brink of depletion. To decelerate the fast depletion of mineral sources of fuels, it is crucial to seek alternative resources that are renewable and sustainable by nature. Significant contribution to pollution reduction can also be observed such as reduction in carbon dioxide emission (CO_2) emission through promotion of renewable energy usage (Dincer & Rosen, 2011). Solar thermal systems have shown tremendous growth in terms of efficiencies but lack the backup to be on the forefront of energy generation despite being a relatively mature technology. Thermal energy storage (TES) systems are commonly employed in construction as well as industrial application for its advantages such as improved overall efficiency, better reliability, etc.

6.4.1 General classification of thermal energy storage system

The thermal energy storage system is categorized under several key parameters such as capacity, power, efficiency, storage period, charge/discharge rate as well as the monetary factor involved. The TES can be categorized into three forms (Khan, Saidur, & Al-Sulaiman, 2017; Sarbu & Sebarchievici, 2018; Sharma, Tyagi, Chen, & Buddhi, 2009):

- Sensible heat storage (SHS)
- Latent heat storage (LHS)
- Thermochemical heat storage

The sensible heat storage is the system of without transformation physical state of materials. But, the latent heat storage system changes the physical state of the materials from solid to liquid or liquid to vapor (Khan et al., 2017). The details thermal energy system and the classification of thermal energy storage materials are shown in Fig. 6.22.

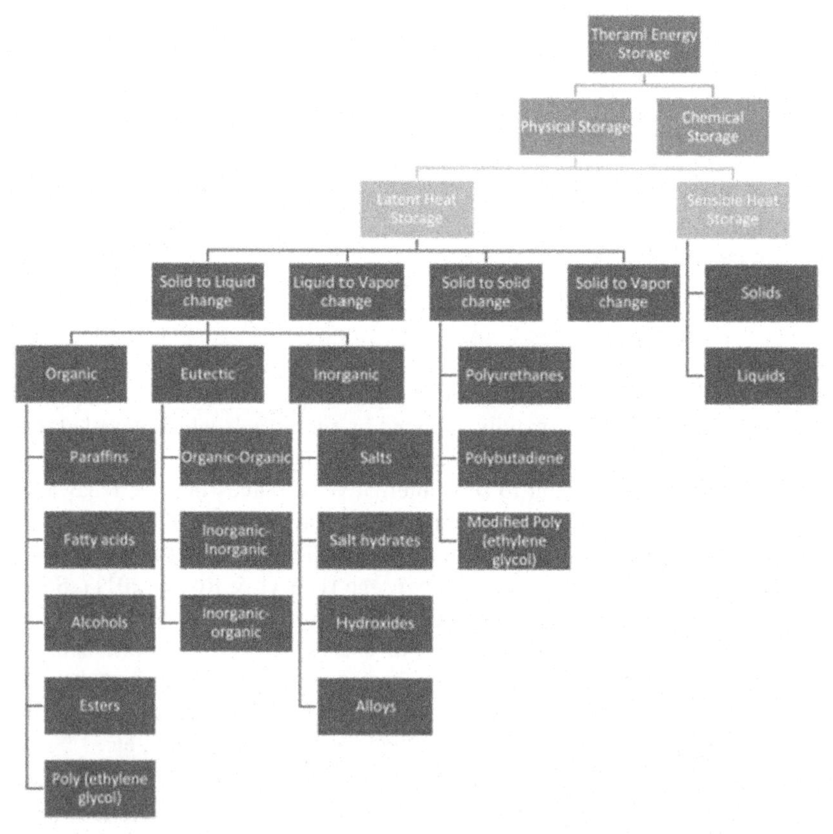

FIGURE 6.22 Details thermal energy system and the classification of thermal energy storage materials (Khan et al., 2017).

6.4.2 Sensible heat storage

Sensible heat storage is system where the energy contained through increasing or decreasing the temperature of storage material. This storage medium is either available as solid or liquid form. Water is one of the most commonly used medium as it is the cheapest. Water is also popular for commercial heat storage with various application in the residential and industrial sector. SHS has two significant benefits in which it's relatively cheap and has minimal danger as lesser harmful materials are used. Fig. 6.23 shows the schematic diagram of water tank storage and the heat source.

The sensible heat storage works based on heat capacity and corresponding temperature variation of respective storage material when the charge and discharge phases take place. Specific heat of the medium, variation in temperature, and the amount of storage material determine how much thermal energy that can be stored. This is described in the following Eq. (6.5) (Kumar & Shukla, 2015):

$$Q_s = \int_{t_i}^{t_f} mc_p dt = mc_p(t_f - t_i) \tag{6.5}$$

where

Q_s is the amount of heat stored (J)
M is the mass of storage medium (kg)
T_i is the initial temperature (°C)
T_f is the final Temperature (°C)
C_p is the specific heat (J/kg·K)

Water is the most commonly used SHS liquid that is available for medium temperature range (0–100 °C) because of its elevated specific heat (4190 J/(kg.K)) and relatively low costing (Ayyappan, Mayilsamy, & Sreenarayanan, 2016). For temperatures greater than 100°C, containment media such as liquid metals, oils, and molten salts are much preferred (Sarbu & Sebarchievici, 2018). Table 6.1 indicates properties of typical solid phase heat containment

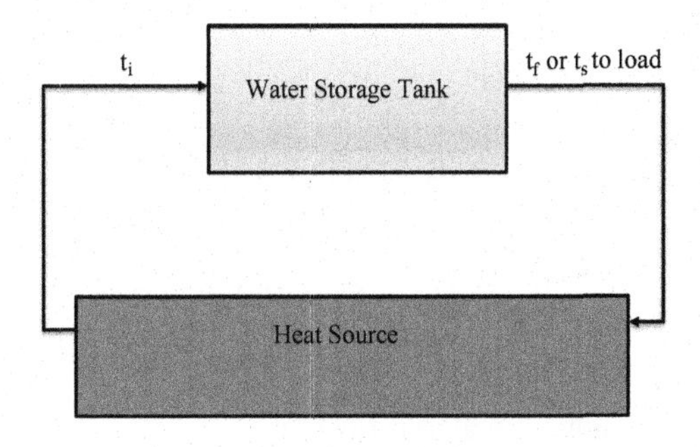

FIGURE 6.23 Schematic diagram of water tank storage.

TABLE 6.1 Solid phase sensible heat storage media (Tian & Zhao, 2013).

Storage materials	Working temperature (°C)	Density (kg/m^3)	Thermal conductivity (W/(m-K))	Specific heat (kJ/(kg.°C))
Sand-rock minerals	200–300	1700	1.0	1.30
Reinforced concrete	200–400	2200	1.5	0.85
Cast iron	200–400	7200	37.0	0.56
NaCl	200–500	2160	7.0	0.85
Cast steel	200–700	7800	40.0	0.60
Silica fire bricks	200–700	1820	1.5	1.00
Magnesia fire bricks	200–1200	3000	5.0	1.15

medium such as ferroalloy, sand-rock minerals, concretes and fire bricks (Tian & Zhao, 2013). These mediums have an excellence thermal conductivity and temperatures range of 200–1200°C.

6.4.2.1 SHS system

Hot-water tank is one of the most commonly used SHS system for conserving energy in water heating system. The SHS Capacity of the water tank can be calculated by using Eq. (6.6) (Sarbu & Sebarchievici, 2018; Sharma et al., 2009):

$$Q_s = mC_p \, \Delta T_s \tag{6.6}$$

where

Q_s is the total heat capacity (J)
m is the mass of storage material (kg)
C_p is the specific heat (J/kg·K)
Δt_s is the temperature range

The tank temperature of the SHS system can be calculated by using the following equation (Sarbu & Sebarchievici, 2018).

$$t_s = t_i + \frac{\Delta\tau}{mc_p[Q_u - Q_L - U_sA_s(t_i - t_a)]} \tag{6.7}$$

where

m is the mass of storage material (kg)
C_p is the specific heat (J/kg·K)

τ is the time
Q_u is the rate of addition
Q_L is the rate of removal
U_s is the heat loss coefficient
A_s is the surface area of storage tank
t_i is the initial temperature
t_a is the ambient temperature

In hot water storage system, charging temperature ranges between 80 and 90°C.

6.4.3 Latent heat storage

Latent heat storage system commonly employs phase change materials (PCMs) during energy containment. PCMs have the potential for energy absorption as well as energy release which involves physical state change. In LHS system, heat containment process takes place during the phase change that occurs over a relatively stable temperature and corresponds proportionally to the fusion heat of substance. The PCMs portrays itself as materials that display elevated energy containment as well as maintaining a uniformed temperature throughout the process. The LHS capacity can be calculated by using the following equation (Sarbu & Sebarchievici, 2018; Sharma et al., 2009):

$$Q_s = m[c_{ps}(t_m - t_i) + f\Delta q + c_{pl}(t_i - t_m) \qquad (6.8)$$

where

Q_s is storage capacity (J)
t_m is the melting temperature (°C)
m is the mass of material (kg)
C_{ps} is the specific heat of solid phase kJ/(kg·K)
C_{pl} is the Specific heat of liquid phase kJ/(kg·K)
f is the melt fraction
Δq is the latent heat of fusion (J)

6.4.3.1 *Phase change materials and characteristics for latent heat storage system*

The PCMs of high thermal conductivity and high heat of fusion are the best options for the LHS system. It should also have a high density and relatively low volume variation when conversion between phases takes places minimizes containment volume, low cost and high availability for large scale applications. The commonly used PCMs for LHS mediums can be widely categorized based on the physical state of transformation for heat absorption and desorption that is shown in Fig. 6.24.

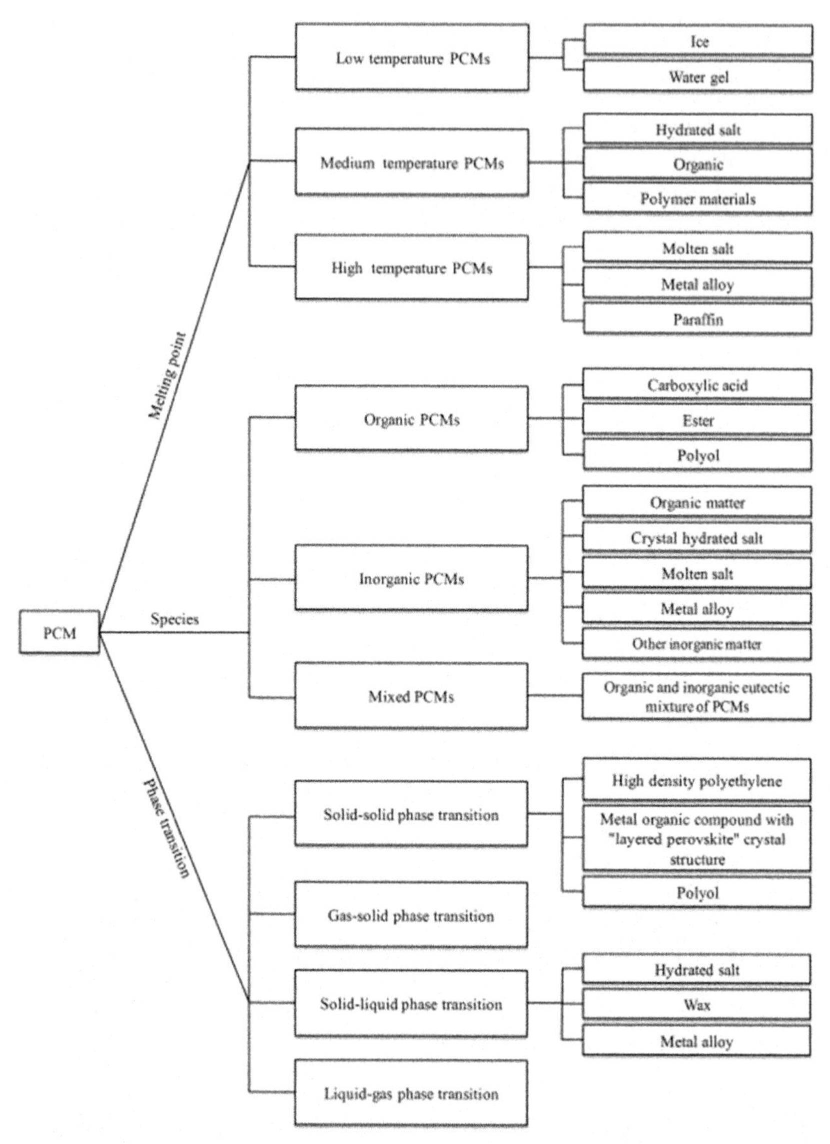

FIGURE 6.24 Classification of PCMs based on different characteristics (Ge, Li, Mei, & Liu, 2013).

Selection of a certain PCM chiefly depends on the specific application. In Table 6.2 some of the most used commercial PCMs are listed. It can be seen from the table that no single PCM possesses all attractive thermal-physical properties. While paraffin is attractive from the viewpoint of latent heat of fusion to meet high melting point requirements, fatty acid and metallic are preferable.

TABLE 6.2 List of some selected phase change materials.

Class	Material	Melting point (°C)	Density (kg/m^3)	Thermal conductivity (W/mK)	Latent heat of fusion (kJ/kg)	Reference
Organic	Paraffin wax	64	916 (solid, 24°C) 769 (liquid, 65°C)	0.346 (solid, 33.6°C) 0.167 (liquid, 63.5°C)	173.6	(Farid, Khudhair, Razack, & Al-Hallaj, 2004; Lane, 1980)
Fatty acid	Palmitic acid	64	850 (at 65°C)	0.162 (at 68.4°C)	185.4	(Farid et al., 2004; Lane, 1980)
Salt hydrate	$CaCl_2.6H_2O$	29	1562 (at 32°C) 1802 (at 24°C)	0.540 (at 38.7°C) 1.008 (at 23°C)	190.8	(Farid et al., 2004; Lane, 1980)
Metallics	Bi-in eutectic	72	n.a.	n.a.	25.0	(Ge et al., 2013)

References

Akhil, A. A., Huff, G., Currier, A. B., Kaun, B. C., Rastler, D. M., Chen, S. B., … Gauntlett, W. D. (2013). *Electricity storage handbook in collaboration with NRECA*. Sandia National Laboratories.

Al Zohbi, G., Hendrick, P., Renier, C., & Bouillard, P. (2016). The contribution of wind-hydro pumped storage systems in meeting Lebanon's electricity demand. *International Journal of Hydrogen Energy, 41*(17), 6996–7004.

Angell, M., Pan, C.-J., Rong, Y., Yuan, C., Lin, M.-C., Hwang, B.-J., & Dai, H. (2017). High Coulombic efficiency aluminum-ion battery using an $AlCl_3$-urea ionic liquid analog electrolyte. *Proceedings of the National Academy of Sciences, 114*(5), 834. https://doi.org/10.1073/pnas.1619795114.

Anne-Chloé Devic, M. I., Suárez, E., Fernández, V., & Bax, L. (2018). *Battery energy storage. SUSCHEM European technology platform for sustainable chemistry*. Retrieved on 02-05-2019 www.suschem.org/files/library/Suschem_energy_storage_final_preview.pdf.

Ayyappan, S., Mayilsamy, K., & Sreenarayanan, V. V. (2016). Performance improvement studies in a solar greenhouse drier using sensible heat storage materials. *Heat and Mass Transfer, 52*(3), 459–467. https://doi.org/10.1007/s00231-015-1568-5.

Bamgbopa, M. O., Almheiri, S., & Sun, H. (2017). Prospects of recently developed membraneless cell designs for redox flow batteries. *Renewable and Sustainable Energy Reviews, 70*, 506–518. https://doi.org/10.1016/j.rser.2016.11.234.

Battke, B., Schmidt, T. S., Grosspietsch, D., & Hoffmann, V. H. (2013). A review and probabilistic model of lifecycle costs of stationary batteries in multiple applications. *Renewable and Sustainable Energy Reviews, 25*, 240–250.

Benato, A., & Stoppato, A. (2018). Pumped thermal electricity storage: A technology overview. *Thermal Science and Engineering Progress, 6*, 301–315.

Bragard, M., Soltau, N., Thomas, S., & De Doncker, R. W. (2010). The balance of renewable sources and user demands in grids: Power electronics for modular battery energy storage systems. *IEEE Transactions on Power Electronics, 25*(12), 3049–3056.

Breeze, P. (2018a). *Hydropower power system energy storage technologies*. Elsevier Science.

Breeze, P. (2018b). *Pumped storage hydropower power system energy storage technologies*. Elsevier Science.

Canales, F. A., & Beluco, A. (2014). Modeling pumped hydro storage with the micropower optimization model (HOMER). *Journal of Renewable and Sustainable Energy, 6*(4), 043131.

Chen, H., Cong, T. N., Yang, W., Tan, C., Li, Y., & Ding, Y. (2009). Progress in electrical energy storage system: A critical review. *Progress in Natural Science, 19*(3), 291–312.

Chen, Y., Freunberger, S. A., Peng, Z., Fontaine, O., & Bruce, P. G. (2013). Charging a Li–O2 battery using a redox mediator. *Nature Chemistry, 5*(6), 489.

Cheng, J., Zhang, L., Yang, Y.-S., Wen, Y.-H., Cao, G.-P., & Wang, X.-D. (2007). Preliminary study of single flow zinc–nickel battery. *Electrochemistry Communications, 9*(11), 2639–2642.

Chen, H., & Lu, Y. C. (2016). A high-energy-density multiple redox semi-solid-liquid flow battery. *Advanced Energy Materials, 6*(8), 1502183.

Council, E. A. S. A. (2017). *Valuing dedicated storage in electricity grids*. EASAC policy report 33.

Decourt, B., & Debarre, R. (2013). *Electricity storage tech. Rep*. Gravenhage, Netherlands: SBC Energy Institute. URL http://energystorage.org/system/files/resources/sbcenergyinstitute_electricity storagefactbook.pdf.

Denholm, P., Jorgenson, J., Hummon, M., Jenkin, T., Palchak, D., Kirby, B., … O'Malley, M. (2013). *Value of energy storage for grid applications*. Golden, CO (United States): National Renewable Energy Lab.(NREL).

Dincer, I., & Rosen, M. A. (2011). *Thermal energy storage: Systems and application*. Chichester, UK: John Wiley & Sons.

Dunn, B., Kamath, H., & Tarascon, J.-M. (2011). Electrical energy storage for the grid: A battery of choices. *Science, 334*(6058), 928–935.

Evers, S., Yim, T., & Nazar, L. F. (2012). Understanding the nature of absorption/adsorption in nanoporous polysulfide sorbents for the Li–S battery. *The Journal of Physical Chemistry C, 116*(37), 19653–19658.

ENER, D. (2019). *Working paper. The future role and challenges of energy storage* [online https://ec.europa.eu/energy/sites/ener/files/energy_storage.pdf (18/07/2019)].

Energy storage – packing some power.(2012). Retrieved from https://www.economist.com/technology-quarterly/2012/03/03/packing-some-power.

Eyer, J., & Corey, G. (2010). Energy storage for the electricity grid: Benefits and market potential assessment guide. *Sandia National Laboratories, 20*(10), 5.

Farid, M. M., Khudhair, A. M., Razack, S. A. K., & Al-Hallaj, S. (2004). A review on phase change energy storage: Materials and applications. *Energy Conversion and Management, 45*(9), 1597–1615. https://doi.org/10.1016/j.enconman.2003.09.015.

Ge, H., Li, H., Mei, S., & Liu, J. (2013). Low melting point liquid metal as a new class of phase change material: An emerging frontier in energy area. *Renewable and Sustainable Energy Reviews, 21*, 331–346. https://doi.org/10.1016/j.rser.2013.01.008.

Guo, R., Zhao, L., & Yue, W. (2015). Assembly of core–shell structured porous carbon–graphene composites as anode materials for lithium-ion batteries. *Electrochimica Acta, 152*, 338–344.

Hadjipaschalis, I., Poullikkas, A., & Efthimiou, V. (2009). Overview of current and future energy storage technologies for electric power applications. *Renewable and Sustainable Energy Reviews, 13*(6–7), 1513–1522.

Hang, B. T., Eashira, M., Watanabe, I., Okada, S., Yamaki, J.-I., Yoon, S.-H., & Mochida, I. (2005). The effect of carbon species on the properties of Fe/C composite for metal–air battery anode. *Journal of Power Sources, 143*(1–2), 256–264.

Hannan, M. A., Hoque, M. M., Mohamed, A., & Ayob, A. (2017). Review of energy storage systems for electric vehicle applications: Issues and challenges. *Renewable and Sustainable Energy Reviews, 69*, 771–789. https://doi.org/10.1016/j.rser.2016.11.171.

Hasanuzzaman, M., Rahim, N. A., Hosenuzzaman, M., Saidur, R., Mahbubul, I. M., & Rashid, M. M. (2012). Energy savings in the combustion based process heating in industrial sector. *Renewable and Sustainable Energy Reviews, 16*(7), 4527–4536. https://doi.org/10.1016/j.rser.2012.05.027.

Hasanuzzaman, M., Zubir, U. S., Ilham, N. I., & Che, H. S. (2017). Global electricity demand, generation, grid system, and renewable energy polices: A review. *Wiley Interdisciplinary Reviews: Energy and Environment, 6*(3). https://doi.org/10.1002/wene.222.

Hazza, A., Pletcher, D., & Wills, R. (2004). A novel flow battery: A lead acid battery based on an electrolyte with soluble lead (II) Part I. Preliminary studies. *Physical Chemistry Chemical Physics, 6*(8), 1773–1778.

Hedegaard, K., & Meibom, P. (2012). Wind power impacts and electricity storage–A time scale perspective. *Renewable Energy, 37*(1), 318–324.

Hoffman, M. G., Kintner-Meyer, M. C., Sadovsky, A., & DeSteese, J. G. (2010). *Analysis tools for sizing and placement of energy storage for grid applications-a literature review: Pacific Northwest National Lab.* Richland, WA (United States): PNNL.

Hossain, F. M., Hasanuzzaman, M., Rahim, N. A., & Ping, H. W. (2015). Impact of renewable energy on rural electrification in Malaysia: A review. *Clean Technologies and Environmental Policy, 17*(4), 859–871. https://doi.org/10.1007/s10098-014-0861-1.

Ibrahim, H., Ilinca, A., & Perron, J. (2008). Energy storage systems—characteristics and comparisons. *Renewable and Sustainable Energy Reviews, 12*(5), 1221–1250.

Imanishi, N., & Yamamoto, O. (2014). Rechargeable lithium–air batteries: Characteristics and prospects. *Materials Today, 17*(1), 24–30.

Jacob, A. (2001). Wind energy—the fuel of the future. *Reinforced Plastics, 45*, 10–13.

Jiao, H., Wang, C., Tu, J., Tian, D., & Jiao, S. (2017). A rechargeable Al-ion battery: Al/molten $AlCl_3$–urea/graphite. *Chemical Communications, 53*(15), 2331–2334.

Jung, S. C., Kang, Y.-J., Yoo, D.-J., Choi, J. W., & Han, Y.-K. (2016). Flexible few-layered graphene for the ultrafast rechargeable aluminum-ion battery. *Journal of Physical Chemistry C, 120*(25), 13384–13389.

Kao, C.-Y., & Chou, K.-S. (2010). Iron/carbon-black composite nanoparticles as an iron electrode material in a paste type rechargeable alkaline battery. *Journal of Power Sources, 195*(8), 2399–2404.

Kapila, S., Oni, A. O., & Kumar, A. (2017). The development of techno-economic models for large-scale energy storage systems. *Energy, 140*, 656–672.

Khalilpour, K. R., & Vassallo, A. (2016). *Community energy networks with storage: Modeling frameworks for distributed generation*. Springer.

Khalilpour, K. R., Vassallo, A. M., & Chapman, A. C. (2017). Does battery storage lead to lower GHG emissions? *The Electricity Journal, 30*(10), 1–7.

Khan, M. M. A., Saidur, R., & Al-Sulaiman, F. A. (2017). A review for phase change materials (PCMs) in solar absorption refrigeration systems. *Renewable and Sustainable Energy Reviews, 76*, 105–137. https://doi.org/10.1016/j.rser.2017.03.070.

Kim, K. J., Park, M.-S., Kim, Y.-J., Kim, J. H., Dou, S. X., & Skyllas-Kazacos, M. (2015). A technology review of electrodes and reaction mechanisms in vanadium redox flow batteries. *Journal of Materials Chemistry A, 3*(33), 16913–16933. https://doi.org/10.1039/C5TA02613J.

Koenig, G. M., Belharouak, I., Deng, H., Sun, Y. K., & Amine, K. (2016). Composition-tailored synthesis of gradient transition metal precursor particles for lithium-ion battery cathode materials. *Chemistry of Materials, 23*, 2863–2870.

Koko, S. P., Kusakana, K., & Vermaak, H. J. (2018). Optimal power dispatch of a grid-interactive micro-hydrokinetic-pumped hydro storage system. *Journal of Energy Storage, 17*, 63–72.

Kolthoff, I., & Tomsicek, W. J. (1935). The oxidation potential of the system potassium ferrocyanide–potassium ferricyanide at various ionic strengths. *The Journal of Physical Chemistry, 39*(7), 945–954.

Koohi-Kamali, S., Tyagi, V., Rahim, N., Panwar, N., & Mokhlis, H. (2013). Emergence of energy storage technologies as the solution for reliable operation of smart power systems: A review. *Renewable and Sustainable Energy Reviews, 25*, 135–165.

Kumar, D., Rajouria, S. K., Kuhar, S. B., & Kanchan, D. K. (2017). Progress and prospects of sodium-sulfur batteries: A review. *Solid State Ionics, 312*, 8–16. https://doi.org/10.1016/j.ssi.2017.10.004.

Kumar, A., & Shukla, S. K. (2015). A review on thermal energy storage unit for solar thermal power plant application. *Energy Procedia, 74*, 462–469. https://doi.org/10.1016/j.egypro.2015.07.728.

Kurzweil, P., & Garche, J. (2017). 2–overview of batteries for future automobiles. In J. Garche, E. Karden, P. T. Moseley, & D. A. J. Rand (Eds.), *Lead-acid batteries for future automobiles* (pp. 27–96). Amsterdam: Elsevier.

Kwon, G., Lee, S., Hwang, J., Shim, H.-S., Lee, B., Lee, M. H., … Kang, K. (2018). Multi-redox molecule for high-energy redox flow batteries. *Joule, 2*(9), 1771–1782. https://doi.org/10.1016/j.joule.2018.05.014.

Lane, G. A. (1980). Low temperature heat storage with phase change materials. *International Journal of Ambient Energy, 1*(3), 155–168. https://doi.org/10.1080/01430750.1980.9675731.

Lemaire, E., Martin, N., Nørgård, P., de Jong, E., de Graaf, R., Groenewegen, J., … Tselepis, S. (2011). *European white book on grid-connected storage*. Savoie Technolac: Department INES Recherche, Développement & Innovation, Institution CEA INES.

Leung, P., Li, X., De León, C. P., Berlouis, L., Low, C. J., & Walsh, F. C. (2012). Progress in redox flow batteries, remaining challenges and their applications in energy storage. *RSC Advances, 2*(27), 10125–10156.

Leung, P., Martin, T., Liras, M., Berenguer, A., Marcilla, R., Shah, A., … Palma, J. (2017). Cyclohexanedione as the negative electrode reaction for aqueous organic redox flow batteries. *Applied Energy, 197*, 318–326.

Leung, P. K., Martin, T., Shah, A. A., Mohamed, M. R., Anderson, M. A., & Palma, J. (2017). Membrane-less hybrid flow battery based on low-cost elements. *Journal of Power Sources, 341*, 36–45. https://doi.org/10.1016/j.jpowsour.2016.11.062.

Li, Y. (2016). A liquid-electrolyte-free anion-exchange membrane direct formate-peroxide fuel cell. *International Journal of Hydrogen Energy, 41*(5), 3600–3604.

Li, Y., & Dai, H. (2014). Recent advances in zinc–air batteries. *Chemical Society Reviews, 43*(15), 5257–5275.

Li, B., & Li, W. (2011). A review of Yangjiang 8th international radiosonde intercomparison. *Advances in Meteorological Science and Technology, 1*, 6–13.

Li, Z., Weng, G., Zou, Q., Cong, G., & Lu, Y.-C. (2016). A high-energy and low-cost polysulfide/iodide redox flow battery. *Nano Energy, 30*, 283–292.

Li, W., Yang, Y., Zhang, G., & Zhang, Y.-W. (2015). Ultrafast and directional diffusion of lithium in phosphorene for high-performance lithium-ion battery. *Nano Letters, 15*(3), 1691–1697.

Liao, K., Mao, P., Li, N., Han, M., Yi, J., He, P., Sun, Y., & Zhou, H. (2016). Stabilization of polysulfides via lithium bonds for Li–S batteries. *Journal of Materials Chemistry A, 4*(15), 5406–5409.

Lin, K., Chen, Q., Gerhardt, M. R., Tong, L., Kim, S. B., Eisenach, L., … Aziz, M. J. (2015). Alkaline quinone flow battery. *Science, 349*(6255), 1529–1532.

Lin, M. C., Gong, M., Lu, B., Wu, Y., Wang, D. Y., Guan, M., … Dai, H. (2015). An ultrafast rechargeable aluminium-ion battery. *Nature, 520*(7547), 325–328. https://doi.org/10.1038/nature14340.

Liu, F.-C., Shadike, Z., Ding, F., Sang, L., & Fu, Z.-W. (2015). Preferential orientation of I_2-LiI $(HPN)_2$ film for a flexible all-solid-state rechargeable lithium–iodine paper battery. *Journal of Power Sources, 274*, 280–285.

Liu, Q., Shinkle, A. A., Li, Y., Monroe, C. W., Thompson, L. T., & Sleightholme, A. E. (2010). Non-aqueous chromium acetylacetonate electrolyte for redox flow batteries. *Electrochemistry Communications, 12*(11), 1634–1637.

Liu, Y., Li, N., Wu, S., Liao, K., Zhu, K., Yi, J., … Zhou, H. (2015). Reducing the charging voltage of a Li–O_2 battery to 1.9 V by incorporating a photocatalyst. *Energy and Environmental Science, 8*(9), 2664–2667.

Luo, X., Wang, J., Dooner, M., & Clarke, J. (2015). Overview of current development in electrical energy storage technologies and the application potential in power system operation. *Applied Energy, 137*, 511–536.

Lu, B., & Shahidehpour, M. (2005). Short-term scheduling of battery in a grid-connected PV/ battery system. *IEEE Transactions on Power Systems, 20*(2), 1053–1061.

Makansi, J., & Abboud, J. (2002). *Energy storage*. Energy Storage Council White Paper.

Manthiram, A., & Yu, X. (2015). Ambient temperature sodium–sulfur batteries. *Small, 11*(18), 2108–2114.

Ma, T., Yang, H., Lu, L., & Peng, J. (2014). Technical feasibility study on a standalone hybrid solar-wind system with pumped hydro storage for a remote island in Hong Kong. *Renewable Energy, 69*, 7–15.

McKerracher, R., Ponce de Leon, C., Wills, R., Shah, A., & Walsh, F. C. (2015). A review of the iron–air secondary battery for energy storage. *ChemPlusChem, 80*(2), 323–335.

Mohamed, M. R., Leung, P. K., & Sulaiman, M. H. (2015). Performance characterization of a vanadium redox flow battery at different operating parameters under a standardized test-bed system. *Applied Energy, 137*, 402–412. https://doi.org/10.1016/j.apenergy.2014.10.042.

Nzotcha, U., Kenfack, J., & Blanche Manjia, M. (2019). Integrated multi-criteria decision making methodology for pumped hydro-energy storage plant site selection from a sustainable development perspective with an application. *Renewable and Sustainable Energy Reviews, 112*, 930–947. https://doi.org/10.1016/j.rser.2019.06.035.

Park, M., Ryu, J., Wang, W., & Cho, J. (2017). Material design and engineering of next-generation flow-battery technologies. *Nature Reviews Materials, 2*(1), 16080.

Pei, P., Wang, K., & Ma, Z. (2014). Technologies for extending zinc–air battery's cyclelife: A review. *Applied Energy, 128*, 315–324. https://doi.org/10.1016/j.apenergy.2014.04.095.

Poulain, A., de Dreuzy, J.-R., & Goderniaux, P. (2018). Pump Hydro Energy Storage systems (PHES) in groundwater flooded quarries. *Journal of Hydrology, 559*, 1002–1012. https://doi.org/10.1016/j.jhydrol.2018.02.025.

Pour, N., Kwabi, D. G., Carney, T., Darling, R. M., Perry, M. L., & Shao-Horn, Y. (2015). Influence of edge- and basal-plane sites on the vanadium redox kinetics for flow batteries. *Journal of Physical Chemistry C, 119*(10), 5311–5318. https://doi.org/10.1021/jp5116806.

Prentis, E. L. (2016). Reconstructing renewable energy: Making wind and solar power dispatchable, reliable and efficient. *International Journal of Energy Economics and Policy, 6*(1), 128–133.

Rastler, D. (2010). *Electricity energy storage technology options: A white paper primer on applications, costs and benefits: Electric power research institute.*

Rehman, S., Al-Hadhrami, L. M., & Alam, M. M. (2015). Pumped hydro energy storage system: A technological review. *Renewable and Sustainable Energy Reviews, 44*, 586–598.

Rohit, A. K., Devi, K. P., & Rangnekar, S. (2017). An overview of energy storage and its importance in Indian renewable energy sector: Part I–Technologies and comparison. *Journal of Energy Storage, 13*, 10–23.

Sarbu, I., & Sebarchievici, C. (2018). A comprehensive review of thermal energy storage. *Sustainability, 10*(1). https://doi.org/10.3390/su10010191.

Schweiss, R. (2015). Influence of bulk fibre properties of PAN-based carbon felts on their performance in vanadium redox flow batteries. *Journal of Power Sources, 278*, 308–313. https://doi.org/10.1016/j.jpowsour.2014.12.081.

Scrosati, B., & Garche, J. (2010). Lithium batteries: Status, prospects and future. *Journal of Power Sources, 195*(9), 2419–2430. https://doi.org/10.1016/j.jpowsour.2009.11.048.

Sharma, A., Tyagi, V. V., Chen, C. R., & Buddhi, D. (2009). Review on thermal energy storage with phase change materials and applications. *Renewable and Sustainable Energy Reviews, 13*(2), 318–345. https://doi.org/10.1016/j.rser.2007.10.005.

Shukla, A., Venugopalan, S., & Hariprakash, B. (2001). Nickel-based rechargeable batteries. *Journal of Power Sources, 100*(1−2), 125−148.

Shyam, B., & Kanakasabapathy, P. (2018). Large scale electrical energy storage systems in India-current status and future prospects. *Journal of Energy Storage, 18*, 112−120.

Steffen, B. (2012). Prospects for pumped-hydro storage in Germany. *Energy Policy, 45*, 420−429.

Strbac, G., Aunedi, M., Pudjianto, D., Djapic, P., Teng, F., Sturt, A., ... Brandon, N. (2012). *Strategic assessment of the role and value of energy storage systems in the UK low carbon energy future*. Report for Carbon Trust(2012).

Tarascon, J.-M., & Armand, M. (2011). *Issues and challenges facing rechargeable lithium batteries Materials for sustainable energy: A Collection of peer-reviewed Research and review Articles from*. Nature Publishing Group. World Scientific.

Tian, Y., & Zhao, C. Y. (2013). A review of solar collectors and thermal energy storage in solar thermal applications. *Applied Energy, 104*, 538−553. https://doi.org/10.1016/j.apenergy.2012.11.051.

Trocino, S., Lo Faro, M., Zignani, S. C., Antonucci, V., & Aricò, A. S. (2019). High performance solid-state iron-air rechargeable ceramic battery operating at intermediate temperatures (500−650 °C). *Applied Energy, 233−234*, 386−394. https://doi.org/10.1016/j.apenergy.2018.10.022.

Untereker, D. F., Schmidt, C.l., Jain, G., Tamirisa, P. A., Hossick-Schott, J., & Viste, M. (2017). 8−power sources and capacitors for pacemakers and implantable cardioverter-defibrillators. In K. A. Ellenbogen, B. L. Wilkoff, G. N. Kay, C.-P. Lau, & A. Auricchio (Eds.), *Clinical cardiac pacing, defibrillation and resynchronization therapy* (5th ed., pp. 251−269). Elsevier.

Vangapally, N., Jindal, S., Gaffoor, S. A., & Martha, S. K. (2018). Titanium dioxide-reduced graphene oxide hybrid as negative electrode additive for high performance lead-acid batteries. *Journal of Energy Storage, 20*, 204−212. https://doi.org/10.1016/j.est.2018.09.015.

Wang, Y., He, P., & Zhou, H. (2012). Li-redox flow batteries based on hybrid electrolytes: At the cross Road between Li-ion and redox flow batteries. *Advanced Energy Materials, 2*(7), 770−779. https://doi.org/10.1002/aenm.201200100.

Wang, Q., Sun, J., Wang, Q., Zhang, D.-a., Xing, L., & Xue, X. (2015). Electrochemical performance of α-MoO_3−In_2 O_3 core−shell nanorods as anode materials for lithium-ion batteries. *Journal of Materials Chemistry A, 3*(9), 5083−5091.

Wänn, A., Leahy, P., Reidy, M., Doyle, S., Dalton, H., & Barry, P. (2012). *Environmental performance of existing energy storage installations: February.*

Wei, L., Wu, M., Zhao, T., Zeng, Y., & Ren, Y. (2018). An aqueous alkaline battery consisting of inexpensive all-iron redox chemistries for large-scale energy storage. *Applied Energy, 215*, 98−105.

Wei, X., Xia, G.-G., Kirby, B., Thomsen, E., Li, B., Nie, Z., ... Wang, W. (2016). An aqueous redox flow battery based on neutral alkali metal ferri/ferrocyanide and polysulfide electrolytes. *Journal of the Electrochemical Society, 163*(1), A5150−A5153.

Wu, F., Lee, J. T., Nitta, N., Kim, H., Borodin, O., & Yushin, G. (2015). Lithium iodide as a promising electrolyte additive for lithium−sulfur batteries: mechanisms of performance enhancement. *Advanced Materials, 27*(1), 101−108.

Wu, M. S., Xu, B., Chen, L. Q., & Ouyang, C. Y. (2016). Geometry and fast diffusion of AlCl4 cluster intercalated in graphite. *Electrochimica Acta, 195*, 158−165. https://doi.org/10.1016/j.electacta.2016.02.144.

Wu, S., Yi, J., Zhu, K., Bai, S., Liu, Y., Qiao, Y., ... Zhou, H. (2017). A super-hydrophobic quasi-solid electrolyte for Li-O_2 battery with improved safety and cycle life in humid atmosphere. *Advanced Energy Materials, 7*(4), 1601759.

XAKTY. (2019). *Electrochemistry: Copper zinc battery.* Available at http://xaktly.com/Electrochemistry.html.

Xia, H., Luo, Z., & Xie, J. (2012). Nanostructured $LiMn_2O_4$ and their composites as high-performance cathodes for lithium-ion batteries. *Progress in Natural Science: Materials International, 22*(6), 572–584. https://doi.org/10.1016/j.pnsc.2012.11.014.

Xiao, J., Mei, D., Li, X., Xu, W., Wang, D., Graff, G. L., ... Aksay, I. A. (2011). Hierarchically porous graphene as a lithium–air battery electrode. *Nano Letters, 11*(11), 5071–5078.

Xin, S., Yin, Y. X., Guo, Y. G., & Wan, L. J. (2014). Batteries: A high-energy room-temperature sodium-sulfur battery (adv. Mater. 8/2014). *Advanced Materials, 26*(8), 1308-1308.

Xu, J., Chen, G., Zhang, H., Zheng, W., & Li, Y. (2015). Electrochemical performance of Zr-doped $Li_3 V_2 (PO_4)_3$/C composite cathode materials for lithium ion batteries. *Journal of Applied Electrochemistry, 45*(2), 123–130.

Yang, N., Fu, Y., Yue, H., Zheng, J., Zhang, X., Yang, C., & Wang, J. (2019). An improved semi-empirical model for thermal analysis of lithium-ion batteries. *Electrochimica Acta, 311*, 8–20. https://doi.org/10.1016/j.electacta.2019.04.129.

Yang, C.-J., & Jackson, R. B. (2011). Opportunities and barriers to pumped-hydro energy storage in the United States. *Renewable and Sustainable Energy Reviews, 15*(1), 839–844.

Yang, X.-G., Ye, Q., Cheng, P., & Zhao, T. S. (2015). Effects of the electric field on ion crossover in vanadium redox flow batteries. *Applied Energy, 145*, 306–319. https://doi.org/10.1016/j.apenergy.2015.02.038.

Ye, B. (2014). Natural herbs against cestodes. In H. Mehlhorn (Ed.), *Encyclopedia of parasitology* (pp. 1–3). Berlin, Heidelberg: Springer Berlin Heidelberg.

Yi, J., Liu, Y., Qiao, Y., He, P., & Zhou, H. (2017). Boosting the cycle life of $Li-O_2$ batteries at elevated temperature by employing a hybrid polymer–ceramic solid electrolyte. *ACS Energy Letters, 2*(6), 1378–1384.

Yu, X., Liu, H., Niu, X., Akhberdi, O., Wei, D., Wang, D., & Zhu, X. (2017). The Gα1-cAMP signaling pathway controls conidiation, development and secondary metabolism in the taxol-producing fungus *Pestalotiopsis microspora. Microbiological Research, 203*, 29–39. https://doi.org/10.1016/j.micres.2017.06.003.

Yu, X., & Manthiram, A. (2015). Na_2S–Carbon nanotube fabric electrodes for room-temperature sodium–sulfur batteries. *Chemistry–A European Journal, 21*(11), 4233–4237.

Zabalawi, S. A., Mandic, G., & Nasiri, A. (2008). Utilizing energy storage with PV for residential and commercial use. In *Paper presented at the 2008 34th annual conference of IEEE industrial electronics*.

Zafirakis, D., Chalvatzis, K. J., Baiocchi, G., & Daskalakis, G. (2013). Modeling of financial incentives for investments in energy storage systems that promote the large-scale integration of wind energy. *Applied Energy, 105*, 138–154.

Zhang, X., Wang, X.-G., Xie, Z., & Zhou, Z. (2016). Recent progress in rechargeable alkali metal–air batteries. *Green Energy and Environment, 1*(1), 4–17. https://doi.org/10.1016/j.gee.2016.04.004.

Zhang, C., Wei, Y.-L., Cao, P.-F., & Lin, M.-C. (2018). Energy storage system: Current studies on batteries and power condition system. *Renewable and Sustainable Energy Reviews, 82*, 3091–3106.

Zhao, Y., & Byon, H. R. (2013). High-performance lithium-iodine flow battery. *Advanced Energy Materials, 3*(12), 1630–1635.

Zhao, Y., Hong, M., Bonnet Mercier, N.g., Yu, G., Choi, H. C., & Byon, H. R. (2014). A 3.5 V lithium–iodine hybrid redox battery with vertically aligned carbon nanotube current collector. *Nano Letters, 14*(2), 1085–1092.

Zhao, Y., Wang, L., & Byon, H. R. (2013). High-performance rechargeable lithium-iodine batteries using triiodide/iodide redox couples in an aqueous cathode. *Nature Communications, 4*, 1896.
Zhou, G., Li, L., Wang, D. W., Shan, X. Y., Pei, S., Li, F., & Cheng, H. M. (2015). A flexible sulfur-graphene-polypropylene separator integrated electrode for advanced Li–S batteries. *Advanced Materials, 27*(4), 641–647.

Chapter 7

Energy economics

Laveet Kumar, M.A.A. Mamun, M. Hasanuzzaman
Higher Institution Centre of Excellence (HICoE), UM Power Energy Dedicated Advanced Centre (UMPEDAC), Level 4, Wisma R&D, University of Malaya, Jalan Pantai Baharu, Kuala Lumpur, Malaysia

7.1 Introduction

Energy economics is a part of applied economics that deals with the basic economic issues of assigning energy resources using economic principles, tools, and business models. Normally, microeconomic aspects of energy demand and supply, macroeconomic aspects of the budgeting and investment of energy projects and energy–economy interactions, and the policy framework of the energy sector are important aspects of energy economics. The main objective of this chapter is to present overview of fundamental economic principles, business models and tools that can be used to investigate the issues during implementation of energy sector projects.

7.2 Cost concept

The cost concept can be defined as the expenses incurred on a definite thing or activity. Cost is defined as the monetary measure of the amount of resources utilized for some specified objective. Many parameters related to costs need to be analyzed when assessing a set of feasible alternatives, such as initial or capital investment, building renovation or new construction, labor, materials and spare parts, training and workshops, hardware, accessories and software, technical support, and general support costs. Costs can be classified in a number of ways: operational and nonoperational costs, direct and indirect costs, fixed and variable costs, etc. Several costs are discussed in the following section.

7.2.1 Fixed and variable costs

Costs that are independent of output are called fixed costs. These are fixed throughout the relevant period, whereas costs that are dependent of output are called variable costs. Fixed and variable costs are shown in Fig. 7.1.

Energy for Sustainable Development. https://doi.org/10.1016/B978-0-12-814645-3.00007-9

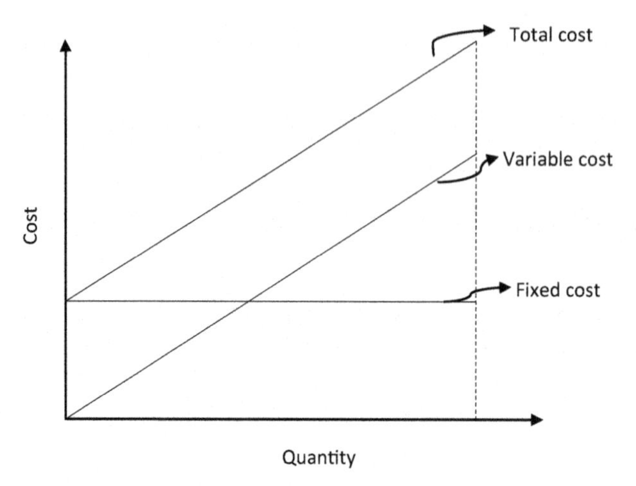

FIGURE 7.1 Fixed and variable costs.

Fixed costs are for the rent of buildings and equipment, salaries of employees, administrative costs, etc. Variable costs, on the other hand, are for raw materials, energy consumption, temporary labor costs, etc. The figure below shows that the number of units or quantity of a product is directly proportional, in rough terms, to total costs—i.e., total costs increase proportionately with increases in product quantity.

Therefore, the sum of fixed and variable costs is total cost.

$$\text{Total Cost} = \text{Fixed Cost} + \text{Variable Cost} \quad (7.1)$$

7.2.2 Breakeven point, profit and loss region

The point when the total cost of the product becomes equal to the revenue made by that product is called the **breakeven point**. Fig. 7.2 shows the

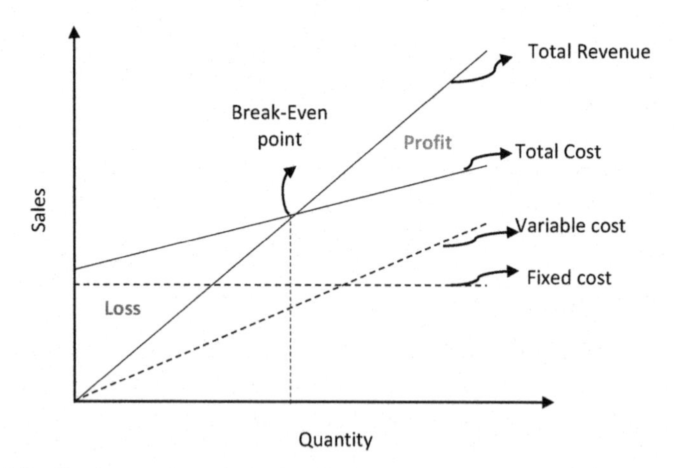

FIGURE 7.2 Breakeven point analysis. *From Barletta, I., Despeisse, M., & Johanssona, B. (2018).*

breakeven point analysis in a measurement system. It is defined as the point in a project where total revenues and total expenses are equal. When the output level of the variable "x" becomes greater than the breakeven point, that region is **the profit region**. Conversely, if the output level of the variable "x" is less than the breakeven point, that region is the **loss region**.

7.3 Money value or time value of money

In financial management, money value or time value of money (TVM) is an important concept. It is the potential earning capacity of some sum of money that can make its value greater in the future than in the present. It is also often known as the present discounted value. Money may be viewed similarly to paying rent on an apartment. The only difference with money is that the rent is known as interest. TVM calculations may change slightly depending on conditions. In general, the most fundamental TVM formula is given as (Investopedia, 2017)

$$FV = PV\left(1 + \left(\frac{i}{n}\right)\right)^{nt} \tag{7.2}$$

where

Future value $= FV$
Present value $= PV$
Interest rate $= i$
Number of compounding periods per year $= n$
Number of years $= t$

7.3.1 Simple interest

Simple interest is defined as the product of the daily interest rate and the total number of days that elapse between payments. The basic formula for calculating interest is given as

$$F_n = P + I_n \tag{7.3}$$

where

$F_n =$ future amount at the end of the nth year
$I_n =$ amount of accumulated interest over n years
$n =$ number of years elapsed between P and F
$P =$ present amount of money at the beginning of the year

7.3.2 Compound interest

Compound interest is a method to calculate interest on the sum of an initial amount plus the accumulated interest on that amount over previous years. The basic formula for calculating compound interest is given as

$$CI = P[(1+i)^n - 1] \tag{7.4}$$

where

CI = compound interest
i = annual interest rate (%)
n = number of compounding years
P = initial amount

In other words, the relationship between a present sum and its equivalent future sum is given as

$$\text{Future sum} = (\text{Present sum})\ (1+i)^n \tag{7.5}$$

7.3.3 Net present value

Net present value (NPV) is defined as the difference between the present value of cash inflows and outflows. If the value obtained through discounted cash flow is higher than the investment cost (i.e., if NPV > 0), the project is feasible. The formula for calculating NPV is given as

$$\text{NPV} = \sum_{t=1}^{T} \frac{C_t}{(1+r)^t} - C_o \tag{7.6}$$

where

Net cash inflows = C_t
Total investment = C_o
Discount rate = r
Number of time periods = t

7.3.4 Internal rate of return

The interest rate at which the NPV of all cash flows from a project equates to zero is known as the internal rate of return. Calculations of IRR are the same as those of NPV.

Furthermore, if the minimum acceptable rate of return (MARR) changes, no new calculations are required, and a decision can be made based on a comparison of IRR and MARR:

The project is attractive if IRR is greater than MARR (IRR > MARR).
The project is unattractive if IRR is less than MARR (IRR < MARR).
The project is neutral if IRR is equal to MARR (IRR = MARR).

7.3.5 Least cost planning (levelized cost)

The most efficient method of performing a given task to reach a specified objective or benefits is the least-cost analysis method. The lowest NPV would be the least cost. The formula for calculating the levelized cost of electricity generation of renewable energy technologies is given as (IRENA, 2012)

$$LCOE = \frac{\sum_{t=1}^{n} \frac{I_t + M_t + F_t}{(1+r)^t}}{\sum_{t=1}^{n} \frac{E_t}{(1+r)^t}} \quad (7.7)$$

where

Electricity generation $= E_t$
Fuel expenditures $= F_t$
Investment expenditures $= I_t$
Operations and maintenance expenditures $= M_t$
Economic life $= n$
Discount rate $= r$

7.3.6 Simple payback period

The basic measure of the financial feasibility of a project is the simple payback period, which is the time required to recuperate the cost of an initial investment. In other words, it can be defined as a simple method to estimate the risk of an investment. Simple payback period can be calculated using the equation given as (Habib, Hasanuzzaman, Hosenuzzaman, Salman, & Mehadi, 2016; Hasanuzzaman, Rahim, Saidur, & Kazi, 2011; Malek, Hasanuzzaman, Rahim, & Al Turki, 2017; Saidur, Hasanuzzaman, Yogeswaran, Mohammed, & Hossain, 2010; Saidur, Rahim, Hasanuzzaman, 2010)

$$PBP = \frac{\text{Incremental cost}}{\text{Annual bill saving}} \quad (7.8)$$

7.4 Business models

Business model structure describes the methods for business investments. It includes design, implementation, management, and critical financing and monitoring features. Ownership and service models are two classifications of business models. Models that emphasize financing and mitigating risks are known as **ownership models**. Models that emphasize facilitating and pointing out different operation and maintenance techniques are known as **service models**. Normally, combining elements of various types and approaches, known as a hybrid approach, is incorporated in most real-world conditions (Rolland, 2011).

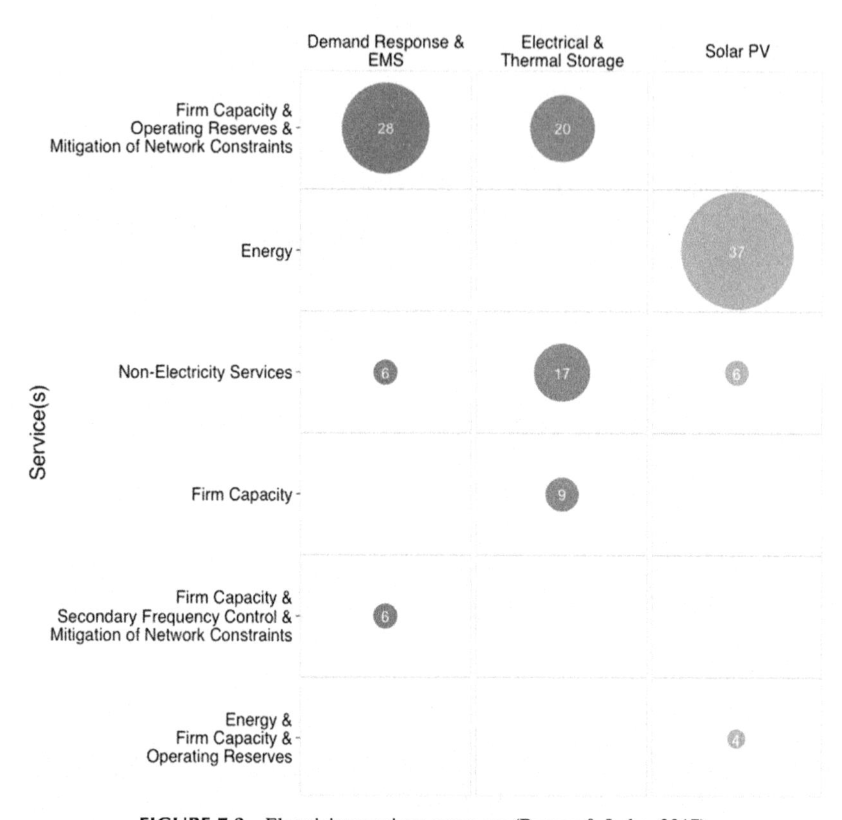

FIGURE 7.3 Electricity services summary (Burger & Luke, 2017).

A summary of various types of business models is given in Fig. 7.3–7.5 (Burger & Luke, 2017). The electricity facilities emphasized by the business models are summarized in Fig. 7.3. The data set considered in this section covers the majority of business models (Masiello, Roberts, & Sloan, 2014; Rebours, Kirschen, Trotignon, & Rossignol, 2007).

Fig. 7.4 is a summary of the customer focuses of various business models. Intermediaries between the two agents are represented by the double-sided arrow (<->) that connects the two customer segments. In the schematic, "C/I/M" is commercial, institutional, or municipal customers. "DER Provider" is the one selling products to businesses. "Regulated Utility" is used to represent the customer segment, which defines that network companies may be involved in transmission, distribution, or both.

Fig. 7.5 summarizes the revenue leveraged by the different business models. The structure of revenue streams depends on the customers targeted.

7.4.1 Demand response and energy management system model

In recent years, increased attention on the business models for DR and EMS have been observed (Behrangrad, 2015; Palensky & Dietrich, 2011). For providing

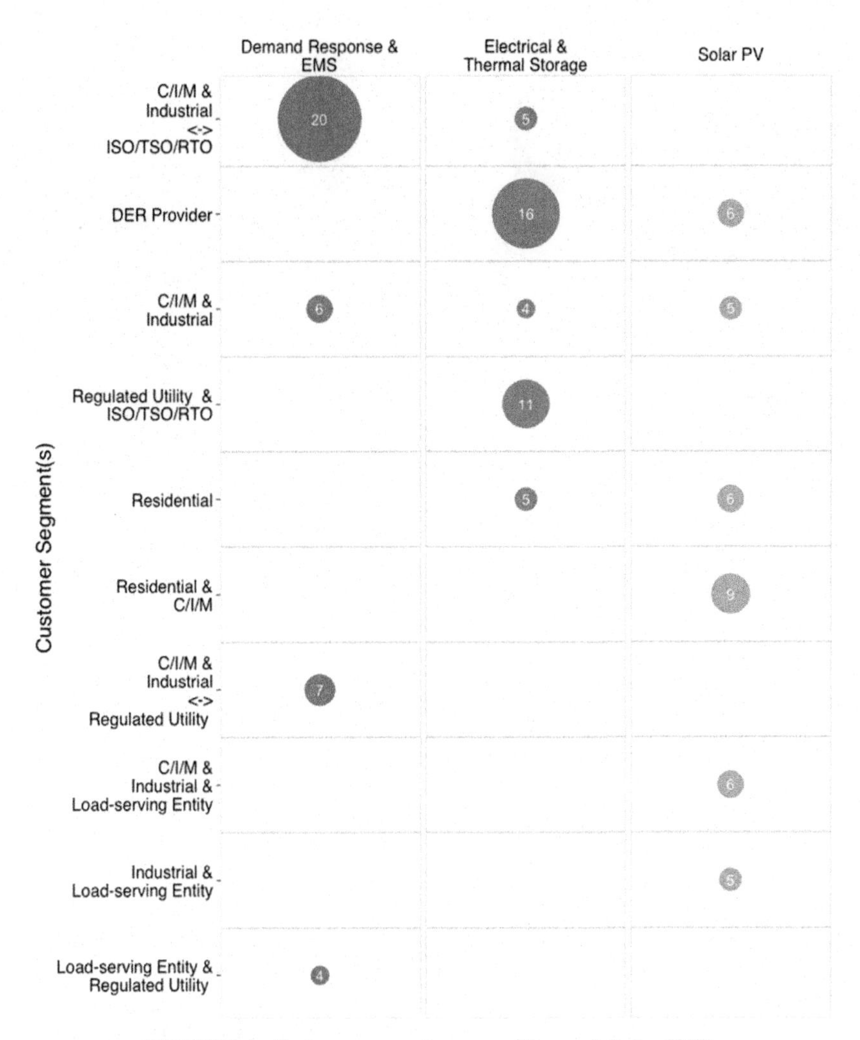

FIGURE 7.4 Customer segment summary (Burger & Luke, 2017).

electricity services, generally energy loads are adjusted, which is known as DR. Automatic activation can be obtained for price signals via an alternative dispatch signal or manually in response to a request. Generally, for monitoring and controlling energy loads, EMS are computer-based models. Based on the method of activation, DR can be classified into three categories: reliability-, economic-, and price-based (Burger & Luke, 2017)

7.4.2 Electricity and thermal storage model

Energy storage devices, either electrical or thermal, are acclaimed as the major technologies of clean energy. Presently, molten salt thermal and pumped hydro

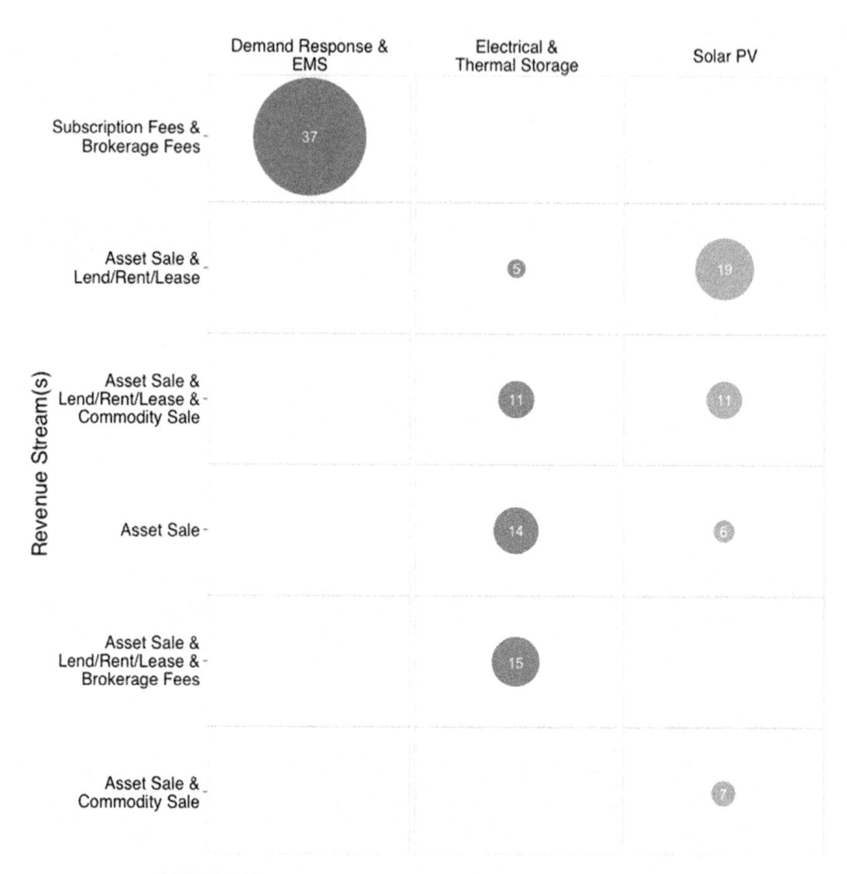

FIGURE 7.5 Revenue streams summary (Burger & Luke, 2017).

energy storage are installed in large capacity but are not suitable for distributed applications (U.S., 2015). Globally, lead-acid storage devices are installed for distributed energy storage applications, although other advanced storage technologies such as lithium-ion (Li-ion) and graphene-based storage technologies are gaining attention (Agnew & Dargusch, 2015). Energy storage devices have a tendency to be relatively modular, integrated, and diverse (Burger & Luke, 2017).

7.4.3 Solar photovoltaic model

Wafer-based and thin film are two major classification of solar photovoltaic technologies (Jean, Brown, Jaffe, Buonassisi, & Bulović, 2015). At present, around 90% of total global production capacity is in the form of wafer-based crystalline silicon (c-Si) modules, while thin film makes up the remainder of global production capacity. Retrospective analysis reveals that the comparatively small distributed generation market is dominated by c-Si modules due to a number of factors that include their higher efficiency.

7.4.4 Ownership business models

Ownership business models describe the capital costs or investment, technical issues, and challenges of energy efficiency and renewable energy. The most suitable ownership business model for grid-connected or medium-to large-scale projects is frequently a public–private partnership (PPP). A build–own–operate–transfer (BOOT) or multiparty ownership system is implemented, for example. A PPP–BOOT business model for a power plant project is illustrated in Fig. 7.6 (ADB, 2015).

7.4.5 Service business models

The service-based business model is used to deliver a product or facilitates the end user, customer, or client by providing services. An energy service company is responsible for providing the services. The energy service company can be a private or public company or a nongovernmental organization.

Energy service companies may be categorized into two groups:

- **Energy supply company**: responsible for supplying electricity, heat, or any other forms of energy to the client.
- **Energy performance company**: responsible for providing energy savings techniques and methods to the client.

For rural electrification, infrastructure and generation assets must first be settled, and for that a user cooperative business model may be formed (ADB, 2015).

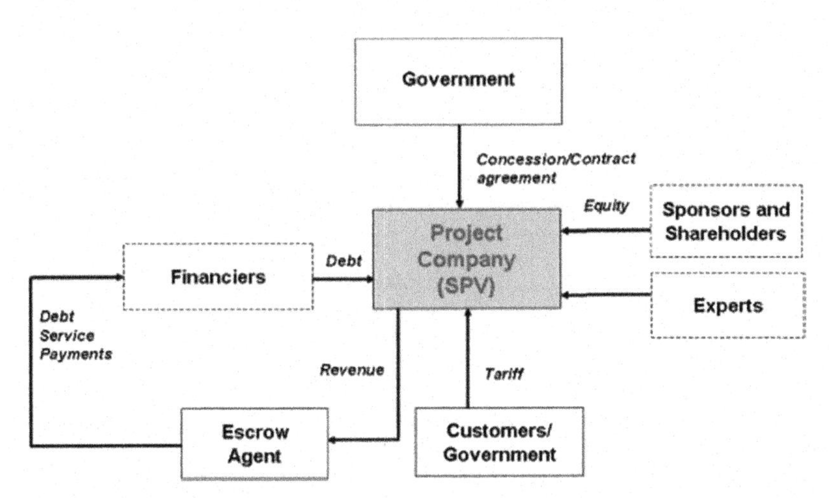

FIGURE 7.6 Relationship diagram of PPP–BOOT model for a power plant project (Investopedia, 2017). *(Source: Chauhan, Y & Marisetty, V. B., 2019).*

7.4.6 Analysis of business models

The analyses of business models are generally on the basis of new revenue models or on a new financing scheme can be are wrapped up with a SWOT analysis, discussion and conclusions for EC models (Würtenberger, Bleyl, Menkveld, Vethman, & van-Tilburg, 2012).

7.5 Barriers to renewable energy

Renewable or sustainable energy has the vast potential to cater to the energy needs of vast populations in developing nations who do not yet have access to sustainable energy. Renewable energy can be an important source for providing a sustainable energy solution to developing nations. Though it is economically feasible for several applications, it has not yet been able to showcase its potential due to several barriers. Many renewable energy experts claim that renewable energy such as wind, solar, and small-scale hydropower are economically not feasible, but at the same time they agree that it is also ideal for rural areas (Painuly, 2001). A framework and key barriers have been defined here for the identification of these barriers. In literature review, several barriers have been defined: cost-effectiveness, technical, environmental, institutional, nonuniform pricing structures, political, regulatory, and social. Some barriers may be linked with a technology, while some may be linked with a region or country (Martinot & McDoom, 1999).

An article by Taylor and Bogach (1997) defines a strategy to accelerate RET development and integration in China based on international assistance. Fig. 7.7 illustrates several levels of barriers. The benefit of analyzing these

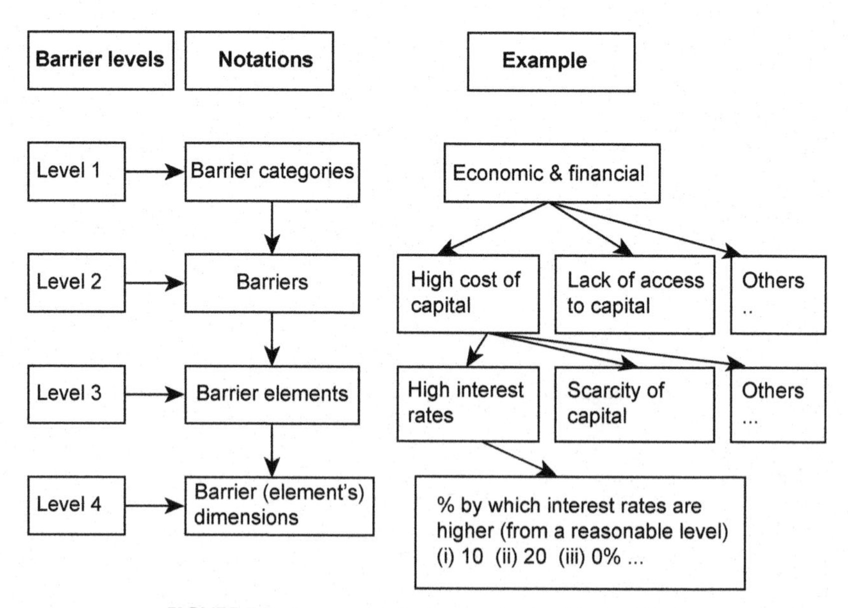

FIGURE 7.7 Framework for barriers levels (Painuly, 2001).

TABLE 7.1 Key barriers (Painuly, 2001).

Category	Barriers
1. Failure of market	Absence of competition and awareness, transaction costs, investment requirements
2. Market distortions	High taxes on renewable energy technologies, subsidies on conventional energy sources
3. Financial	Huge capital cost, discount rates, payback period, initial up-front costs
4. Institutional	Nonexistence of regulatory framework, involvement of stakeholders, and private/professional sector participation
5. Technical	Nonexistence of standard and codes, skilled training facilities, and O & M facilities
6. Social and behavioral	Absence of consumer or client acceptance

barriers on several levels is to highlight the cause for the occurrence of a barrier and therefore may make it easier to respond and address or overcome that barrier. Key barriers have been listed in Table 7.1.

References

ADB. (2015). *Business models to realize the potential of renewable energy and energy efficiency in the greater mekong subregion* (Vol. 6). ADB Avenue, Mandaluyong City, 1550 Metro Manila, Philippines: Retrieved from Asian Development Bank.

Agnew, S., & Dargusch, P. (2015). Effect of residential solar and storage on centralized electricity supply systems. *Nature Climate Change, 5*(4), 315–318.

Barletta, I., Despeisse, M., & Johanssona, B. (2018). The Proposal of an Environmental Break-Even Point as Assessment Method of Product-Service Systems for Circular Economy Procedia CIRP. *72*, 720–725.

Behrangrad, M. (2015). A review of demand side management business models in the electricity market. *Renewable and Sustainable Energy Reviews, 47*, 270–283.

Burger, S. P., & Luke, M. (2017). Business models for distributed energy resources: A review and empirical analysis. *Energy Policy, 109*, 230–248. https://doi.org/10.1016/j.enpol.2017.07.007.

Habib, M. A., Hasanuzzaman, M., Hosenuzzaman, M., Salman, A., & Mehadi, M. R. (2016). Energy consumption, energy saving and emission reduction of a garment industrial building in Bangladesh. *Energy, 112*, 91–100.

Hasanuzzaman, M., Rahim, N. A., Saidur, R., & Kazi, S. N. (2011). Energy savings and emissions reductions for rewinding and replacement of industrial motor. *Energy, 36*(1), 233–240. https://doi.org/10.1016/j.energy.2010.10.046.

Investopedia. (2017). *Time value of money - TVM*. https://www.investopedia.com/terms/t/timevalueofmoney.asp.

IRENA. (2012). Renewable energy technologies: Cost analysis series. *Solar Photovoltaics, 1*(4/5). Power Sector.

Jean, J., Brown, P. R., Jaffe, R. L., Buonassisi, T., & Bulović, V. (2015). Pathways for solar photovoltaics. *Energy and Environmental Science, 8*(4), 1200–1219.

Malek, A. B. M. A., Hasanuzzaman, M., Rahim, N. A., & Al Turki, Y. A. (2017). Techno-economic analysis and environmental impact assessment of a 10 MW biomass-based power plant in Malaysia. *Journal of Cleaner Production, 141*, 502–513. https://doi.org/10.1016/j.jclepro.2016.09.057.

Martinot, E., & McDoom, O. (1999). *Promoting energy efficiency and renewable energy: GEF climate change projects and impacts*. http://www.gefweb.org/PUBLIC/gefcc15.pdf.

Masiello, R. D., Roberts, B., & Sloan, T. (2014). Business models for deploying and operating energy storage and risk mitigation aspects. *Proceedings of the IEEE, 102*(7), 1052–1064.

Painuly, J. P. (2001). Barriers to renewable energy penetration; a framework for analysis. *Renewable Energy, 24*(1), 73–89.

Palensky, P., & Dietrich, D. (2011). Demand side management: Demand response, intelligent energy systems, and smart loads. *IEEE transactions on industrial informatics, 7*(3), 381–388.

Rebours, Y. G., Kirschen, D. S., Trotignon, M., & Rossignol, S. (2007). A survey of frequency and voltage control ancillary services—Part II: Economic features. *IEEE Transactions on Power Systems, 22*(1), 358–366.

Rolland, S. (2011). *Rural electrification with renewable energy*. Brussels: Alliance for Rural Electrification.

Saidur, R., Hasanuzzaman, M., Yogeswaran, S., Mohammed, H. A., & Hossain, M. S. (2010). An end-use energy analysis in a Malaysian public hospital. *Energy, 35*(12), 4780–4785. https://doi.org/10.1016/j.energy.2010.09.012.

Saidur, R., Rahim, N. A., & Hasanuzzaman, M. (2010). A review on compressed-air energy use and energy savings. *Renewable and Sustainable Energy Reviews, 14*(4), 1135–1153. https://doi.org/10.1016/j.rser.2009.11.013.

Taylor, R., & Bogach, V. (1997). *China; a strategy for international assistance to accelerate renewable energy development*. World Bank discussion paper number 388 http://www.worldbank.org/astae/388wbdp.pdf.

U.S. (2015). *DOE. DOE global energy storage database*. Energy Storage Exch 2015 http://www.energystorageexchange.org/.

Würtenberger, L., Bleyl, J. W., Menkveld, M., Vethman, P., & van-Tilburg, X. (2012). *Business models for renewable energy in the built environment*. IEA-RETD/ECN-E–12-014.

Yogesh Chauhan, & Vijaya B. Marisetty. Do public-private partnerships benefit private sector? Evidence from an emerging market. Research in International Business and Finance. Vol. *47*(C), 563–579.

Chapter 8

World energy policies

Nur Iqtiyani Ilham[a,b], M. Hasanuzzaman[a], M.A.A. Mamun[a]
[a]*Higher Institution Centre of Excellence (HICoE), UM Power Energy Dedicated Advanced Centre (UMPEDAC), Level 4, Wisma R&D, University of Malaya, Jalan Pantai Baharu, Kuala Lumpur, Malaysia;* [b]*Faculty of Electrical Engineering, Universiti Teknologi MARA (UiTM), Masai, Johor, Malaysia*

8.1 Introduction

Energy is an indispensable prerequisite for performing work. The availability of energy is an important driver for economic prosperity. The global industrial revolution has contributed to huge developments with social implications and economic growth. For decades, coal, oil, and natural gas have been the energy resources provisioned as fuels for energy. The crude price of fossil fuels today might cause difficulties in meeting rising demand in coming years while the cost of exploiting them will become higher as resources are depleted and the cost to extract them increases (Ritchie and Roser, 2018) (Hasanuzzaman et al., 2012). Furthermore, issues regarding the global climate have led to increased attention by many countries on policy research especially regarding the aspects of energy sustainability and uses of natural resources (Kester, Moyer, & Song, 2015).

Energy issues and their respective research are imperative in developing world and social sciences (Sovacool, 2014). Government at its level must intervene and regulate policies when needed based on valid and reliable evidence obtained from current economic conditions, energy activities and issues, and public and private sector feedback. The Conference of the Parties climate summit agreement (COP21) has turned the energy policy more relevant and well diffusion on its goals through it global scale of policy understanding (Kinley, 2017).

The chapter is divided into two main sections. First, the evaluation of energy policy is reviewed generally based on the dissemination of global energy policy mainly in America, Europe, Asia, Australia, and Africa. In the section thereafter, the roles of government at the federal level in helping energy policy to flourish is discussed, including several instruments such as subsidies, financial incentives, and tax exemptions. This research offers holistic insights into designing energy policy together with its structure and goals.

Energy for Sustainable Development. https://doi.org/10.1016/B978-0-12-814645-3.00008-0

8.2 Mapping global energy policies

Every country of the world has different policies based on its energy markets and its overall economy. The policies formed should considered to reflect respective national prospects and political goals. To ensure the continuous deliverable of energy policy, reference guidelines are essential. Everything related to energy production, distribution, and consumption are to be determined by government policy through energy conservation guidelines, incentives, international agreements, legislation, subsidies, and taxation. COP21, in November 2016, was an initiative taken by most countries globally to take control of issues concerning climate change and global warming. The 195 members of the United Nations Framework Convention on Climate Change (UNFCCC) agreed with COP21 in achieving their mission of achieving an average global temperature below 1.5°C (UN, 2015a; Buxton, 2016; Morgan, 2016).

The awareness about having sustainable energy, clean technologies, and climate changes was a main objective discussed during the "Sustainable Development Round Table" by the UN in 2017. This part assesses energy policy trends globally which relate to the initiatives offered by the respective government and the impacts of policy to accelerate RE activities. Reforming the energy policy entail the information relating to energy efficiency, energy sources, infrastructure and technologies availability and several aspects concerning legislative, political, financial, environmental and climate change (Goldthau, 2016). The policy formed has to be ensured that it can benefited to the technology driven and consumer's privacy security, consumer awareness, transparency in electrical usage and pricing, incentive awarded, technology maturity, adequate funding and market competitiveness TIA (2011). Therefore, this section will elaborate more on the energy policies applied in South America, Europe and Asia in general for comparison and added values.

8.2.1 America

Toward achieving the mission of low carbon emission and overwhelmed the climate crisis, investigation on the impact of energy production/used is essential. In order to overcome with these challenges, the policy set by the government must be formed carefully. The US government, for instance, at its federal and state stages, is responsible for formulating energy policy that benefits all multilayered stakeholders. Known as a noteworthy country in RE investment and its installed capacity, its appointed government and independent agencies such as the Department of Energy, Energy Information Administration (EIA), Environmental Protection Agency, and US Geological Survey are responsible for steering US energy policy (Webber, 2016).

In recent years, Canada has made improvements in developing technologies and equipment that take advantage of the natural resources available in the country. The renewable elements are useable for energy production in the form of electricity, industrial heat, thermal energy for space and water conditioning, and transportation fuels. Realizing the aging of the present generating facilities, an advanced electricity infrastructure is required to meet demand growth and furnish the life span of electricity. The unique characteristics of its geography have expanded the potential of adopting RES for electricity production. Sources such as wind, biomass, solar, geothermal, hydro, and ocean energy are widely used for electricity production. Until now, hydropower and biomass have contributed an enormous portion of kWh production in most Canadian provinces (Barrington-Leigh & Ouliaris, 2017). Canada. through its Mid-Century Long-Term Low-Greenhouse Gas Development Strategy, aimed at electricity production that will be completely originated from RES and the pioneering of CO_2 emission devaluation by 2030. To support the stated policies, the government of Canada has increased the price for CO_2 emissions from CAD10 per tonne of CO_2 (tCO_2) in 2018 to CAD 50/tCO_2 in 2022 (IEA, 2017a).

However, in adopting the innovative energy structure and policy, extensive challenges need to be tackled for mission realization. Significant challenges have been highlighted such as RES intermittence behavior, energy storage, and uneven energy provision at different provinces. Proactive efforts and initiatives should be taken by the government, and analysis on its deployment should be compiled tandemly with review of policy implications.

8.2.2 Europe

For a decade, RE has shown growth in development globally, and the EU is always a leader in promoting RE globally and within the Europe, Middle-East and Africa (EMEA) region. Each country in the EU is responsible for encouraging and managing its own RE sources (RES) and energy mix activities. Despite various challenges that have occurred in aspects of technology maturity and availability, incentives, and consumer engagement at distinctive countries, EU members should directly follow the path of RE policy formed by the European Commission (EC) (IqtiyaniIlham, Hasanuzzaman, & Hosenuzzaman, 2017). The inauguration of the RE directive in EU was written in 2001 and known as Directive 2001/77/CE, with further enhancements made in 2008 under Directive 2009/28/CE. Between those years, the EU launched the 2020 Climate and Energy Package that aimed to enact a 20% increase in energy generated from RES, a 20% cut of greenhouse gas (GHG) emission, and a 20% increase in energy efficiency (EC, 2009). Paving the path of the 2030 Agenda for Sustainable Development, EU development policy has revised the package, targeting 27% RES generated for energy (EC, 2015; UN, 2015b). Indeed, this implies challenges for EU countries in deployment, whereby support from governments in aspects of incentives and specific forces are crucial.

In 2015, the EC initiated the Energy Union Strategy, which is responsible for providing competitive, sustainable, secure, and affordable energy to EU members. Through the path, the EC has underlined five (5) elements on the policy area that can spur the said strategy especially on the energy efficiency, climate actions, research, energy security and European integrated energy market (EC, 2017). Furthermore, energy policy developments in most EU countries are governed by the Paris agreement of 2016.

Germany, for instance, executed the Renewable Energies Heat Act in May 2011 that applied to residential and nonresidential buildings as well as public authorities' existing buildings. The act is initiated from Directive 2009/28/EC on Promoting the Use of Energy from Renewable Sources. The act stated that landowner is obliged to have energy mix originated from RES for heat supply at their respective buildings. The use of solar thermal systems, biomass (solid, liquid, or gas), and geothermal energy/environmental heat is permitted. The owner of the building is free to decide which form of RE (or permitted alternative measures) to use. In France, the government has taken the proactive step by increasing the price for carbon tax (fossil fuels) from EUR 22/tCO_2 (year 2016) to EUR 100/tCO_2 (year 2022). The French and UK governments also announced their 5-year plan which is to exclude the usage of diesel and petrol vehicles on the road by 2040 (UN, 2015a).

Other countries in Europe such as Finland, Sweden, Austria and Denmark listed as top countries in RES integration for the electricity supply. The said governments are very committed concerning climate mitigation. In the year of 2045, Sweden is aiming to become carbon neutral country. While in Finland, the share and usage of RES for final energy consumption has reached 38.7% in 2014, above their target of 38% by 2020, and thus new target of 50% by 2020 was introduced. Although Netherland is well known of its power wind, however it was meant mostly to grin grain and drain water. A report by Eurostat revealed that Netherland is lack behind toward EU 20-20-20 mission whereby until 2016, the share of RES recorded was only 6% (Eurostat, 2018). The main renewable sources in the country are biomass, wind, solar and both geothermal and aerothermal power. Alike Germany, the Dutch governments have been actively adopting various policies and financial incentives to promote the growth of RE within her realm.

Therefore, in order to breed the trust and motivate the stakeholders toward the aims of EU energy policy, it is imperative to ensure the transparency and democratically of the policies formed for the further impact evaluation. With total of 30 countries in the EU organization, thus required the synchronization of the policy. Each member possesses their own national policies which can lead to the issues regarding energy market competitiveness amongst the EU members (Faure-Schuyer, Welsch, & Pye, 2017). Despite various policies possessed by member countries, the policies formed must be coherent and have the same mission as the EU climate and energy package. Further implication on policy will be discussed in detail in the next section—i.e., Government Roles in Perspective.

8.2.3 Asia

Due to rapid revolution in the economic growth and industrial development, Asia is currently encountered with the increasing of energy demand with most of the members states are burgeoning their energy consumption. For instance, China has been acknowledged as the largest energy consumer globally and followed by India due to upsurge of its energy demand with approximately 42% between year 2000 and 2015 (Kumar Singh, 2013; Sahoo et al., 2016). Other member that are also experiencing the increasing of energy demand and consumption listed as Malaysia, Singapore (Ilham et al., 2018), Indonesia (Shahbaz et al., 2013), Vietnam (Tang, Tan, & Ozturk, 2016) and South Korea (Park and Hong, 2013). Japan's energy supply and demand however recorded slightly decreasing by 0.1% in 2017 due to its continue efforts promoting on the energy efficiency. Nevertheless, the efforts do not affect the growth of Japan economic from its macro perspective (Momokoet al., 2017).

Principally, the goals for energy policy in Asia are not deviated much with other continents, which is heightened on sustainable development and climate changes matters. The respective government of the member states is fully responsible in controlling their energy sector that suited with their proposed energy policies. Indonesia for an example is adopted a National Energy Plan as a government regulation that focusing on increasing on the usage of RE, energy mix, energy security and energy access (IEA, 2014). However, it was observed the policy is given less devotion in reducing the CO_2 emission whereby the country is expected doubling the usage of coal and gas. Another example draws is the enacted energy policy in South Korea. The Ministry of Trade, Industry and Energy is accountable for accomplishing the 2035 South Korean New and Renewable Energies (NRE) target. The action plan stated that 11% of energy supply must be originated from RE (IEA, 2011). Hence, the preliminary action toward NRE was taken in April 2011 stated that 10% from the expected total energy of public buildings must be produced by RE (Corporation, 2011).

Known as the "world engine" from the aspect of economics, issues pertaining to sustainable energy and CO_2 emissions should be intensively analyzed. Like other continents, Asia is denoted as a major contributor to global climate change. These energy challenges would entail meticulous study on the environmental, social impacts and energy infrastructure development. To overcome those said climate challenges, China has emphasized several emissions strategies such as energy supply and consumption revolution strategy, ETS, and the 13th Five-Year Plan (2016–20), respectively (Swartz, 2016; IEA, 2017b; The 13th five-year plan of China, 2016). The concept of Japan's energy policy may be adopted to most countries in Asia Pacific, whereby issues concerning energy security, efficiency, and environmental soundness are completely overseen. Roundtable discussion such as the Asia–Europe Energy Policy forum hosted by South Korea in 2017 are essential for discussing in more depth about low-carbon technologies, government roles and incentives, energy policy, and climate mitigation strategies between Asia and Europe Foundation, (2017).

8.2.4 Australia

The key data obtained from the International Energy Association (IEA) revealed that through 2016, approximately two-thirds of electricity generation was produced from coal-fired power. Australia is still dependent on the primary sources of coal and natural gas for energy production, with a major share of total final energy consumption originated from the industrial and transportation sectors (IEA, 2018). To capture the global market trend, Australia is currently enduring with thoughtful transformation, especially on energy and climate policies and its energy market. Initiated from the Energy White Paper, Paris Agreement pledge in 2015, Finkel review, and the 2016 South Australia systemwide blackout, the National Energy Productivity Plan 2015–30 lists several ambitious targets for energy and climate policies by 2030 including a 28% emission reduction of GHGs, 40% energy production enhancement, 33,000 GWh Renewable Energy Target, an Emissions Reduction Fund, and technology development and deployment support, Australian Department of Industry and Science (2015).

The Department of the Environment and Energy (DoEE) has been appointed to oversee several agencies on clean energy activities including RE activities among its states. Initiated from various acts, namely the Australian Renewable Energy Agency Act 2011, Clean Energy Finance Corporation Act 2012, Clean Energy Regulator Act 2011, and Climate Change Authority Act 2011, several government agencies have been established accordingly, following by the Australian Renewable Energy Agency, the Clean Energy Finance Corporation, the Clean Energy Regulator, and the Climate Change Authority, respectively (IEA, 2018). DoEE is also responsible in managing energy data and analysis for all Australian states and territories.

Since 2010, due to the high emissions of GHGs originated from its coal industry, Australia has deployed ETS—a cap-and-trade system to control carbon pollution (Australia, 2010; Pearse, 2016). The Federal Government will initially set the limit (cap) on allowable GHG emitted by the firm's participants based on their nature of works. Any insufficient of permits may require participants to follow several options i.e., reduce the emission below the cap, credit purchasing on the carbon or buying a permit from other firms/overseas market. A positive achievement has been recorded at the earlier year of ETS implementation with 7.4% reduction in carbon pollution, 12% reduction on the usage of coal as generation source and 30% increase of the RES usage (DIICCSRTE, 2013). Although solar and wind power had entered the energy market since 2009, adopting subsidies of those renewable elements such as FiTs is much differing between states in Australia due to some deficiency on the regulations at the Federal side (Poruschi, Ambrey, & Smart, 2018). Assessment made by the IEA suggesting the Council of Australian Governments must develop a set of national principles in the aspect of FiTs amount paid amongst the state and territories.

8.2.5 Africa

Energy poverty is one of the major issues in Africa where about 30% of the population do not have access to electricity. Many people of the countries still use traditional energy for their daily lives such as cooking, lighting, etc. the world Energy council has revealed that energy efficiency, resources used for RE, energy policy and pricing are seen also as issues. Lack of proper planning in the power sector has high loses (about 18%) for transmission and distribution (Avila et al., 2017). As there are many challenges, so many ways and policies need to be revealed for the energy sector. New energy resources, new and cheaper technologies, new policies, financial incentives etc. need to be implemented for the future energy access all people. Off-grid and mini-grid initiatives are the most potential option of many countries in Africa. As a result, mainly changing in especially rural communities, many of them access electricity for the first time in their lives (Africa, 2018).

8.3 Government roles in perspective: initiatives and impacts of energy policy

8.3.1 Subsidies for renewable energy

The world is currently having major transitions within the landscape of energy system. The global shift has happened because of the rapid growth in energy demand and falling costs for RE, China energy revolution and US shale-gas and tight-oil revolution (IEA, 2017a). Despite of having difference preferences in flourishing the energy supply and services, each region have the mutual aimed toward the climate issues, energy access and energy efficiency. The determination of RE subsidies is to preserve the prices that benefit RE consumers and power producers. The said subsidies can be described, such as monetary incentives, tax exemptions, and tax rebates. The share for RE can be designated into the power sector, transportation, and heating and cooling. Several factors are contributing to the growth of RE deployment globally including: tremendous awareness regarding energy security, increase in energy demand, policy and subsidies initiatives, and environment and human health issues.

Many countries have signaled their intention to shift from fossil fuel generation into large-scale RE. For example, in promoting the investment and practice in RE, the Chinese central government set considerable subsidies for this industry, and most of them come from two sources: (1) surcharges on electricity consumption and (2) the National Renewable Energy Development Fund (RED fund). The surcharge is imposed by the central government on every kWh of electricity consumption, with exemption on the residential and agricultural sectors. The surcharge collected is allocated to the grid to purchase renewable electricity and the surplus is deposited into the RED fund. In cases of gap between the annual subsidy and the taxed surcharge, RED fund will be initiated to cover the gap. The said surcharge collection method was also been applied in some Asian countries, such as Thailand and India.

While in Europe, members are driven through the EU 20-20-20 mission. The Dutch government, for example, has committed to a legally binding target of achieving 14% of final energy consumption from RE sources by 2020. Nevertheless, the Dutch government took up a more ambitious challenge by increasing this figure to 16% by 2023. To achieve the figure determined, it is inevitable that a major enhancement in the RE subsidy schemes is required. Since 2003, three subsidies, namely the Environmental Quality of the Electricity Production, Stimulation of Sustainable Energy Production (SDE), and SDE+, have been introduced. Those schemes are carrying its function to passage the subsidy for generators of electricity from renewable sources, subsidizes the compensation mechanism between electricity production by fossil fuel and electricity production with renewables, and as a feed-in premium for SDE respectively (Niesten, Jolink, & Chappin 2018). To finance these policies, the Dutch government impose taxes on the energy bill of end consumers.

8.3.2 Feed-in tariffs and net energy metering

Feed-in tariffs (FiTs) and net energy metering (NEM) are policy frameworks used to encourage and accelerate RE investment and technologies while at the same time achieving the global decarbonized percentage target. This is achievable by permitting RE producers/investors to be remunerated for any kWh energy they feed back into the electricity grid. Both policy frameworks are being applied worldwide, but which one is implemented initially and how it is being executed depends on the respective government. In Malaysia, between 2011 and 2016, the FiT mechanism was implemented before being replaced by the deployment of NEM (SEDA, 2016). While FiTs do not allow producers to use the clean energy generated, since 100% of the energy generated must be supplied to the grid, NEM is applying the concept of "Clean Energy Cashback" that has taken the initiatives of saving electricity bills while making profits. Globally, due to high penetration of FiT, purchasing price is decreasing every year (Pyrgou, Kylili, & Fokaides, 2016). This affects Europe, Asia, and some countries in North America.

A FiT is an energy supply policy that promotes the rapid deployment of RES. The implementation of FiT was first introduced globally during the 1990s and 2000s. FiT can be classified as a gross FiT or a net FiT, and both are suitable for implementation at the national, state, or regional level (Poruschi et al., 2018). Basically, FiT offers a guarantee of payments to RE developers, such as those for solar PV, wind power, hydropower and biomass, for the electricity they produce. These payments are generally awarded as long-term contracts set over a period of 15–20 years. However, the payment structures for FiT are differ subject to market saturation ant policy engaged at the respective countries either to follow with fixed price or premium price payments. Most of the countries in Europe, such as the Netherlands, Germany, and France, choose to adopt fixed price payments for their FiT activities (Toby et al., 2010). Currently, six US states (California, Hawaii, Maine, Oregon, Vermont, and Washington) mandate FITs or similar programs. A few other states have utilities with voluntary FITs. There is growing interest in FiT programs in the United States, especially as evidence mounts about their effectiveness as a framework for promoting RE development and job creation.

To date, FIT policies are successful around the world, notably in Germany. Known as a key player in RE development and deployment, Germany possesses higher RE contribution from solar, biomass, geothermal, wind, and hydro power. Early on in promoting FiT, Strom ein speisungs gesetz (StrEG) (Electricity Feed Act) and Erneuerbare-Energien-Gesetz (EEG) (Renewable Energy Sources Act) has initiated several of favorable incentives to RE investors. StrEG has established a satisfying tariff rates for wind power generation and biomass power generation. Additional improvement was made in 2000 through EEG whereby a fixed tariff calculation method was implemented according to the type of RE generated and share of capacity replacing previous percentage-based tariff rate calculation method introduced by StrEG (Mendonça and Corre, 2009). The purchasing term for FiT sign are 20 years whereby the RE investor is obliged to supply the electricity to grid.

Earlier of FiT implementation, many governments offered high tariff rates for electricity buying. This is to attract RE investor for participating and adoption of RE. Started implementing FiT in 2012, Japan was known as the country that offered highest tariff for RE investor to encourage the deployment of RE[1] (Chen, Kim, & Yamaguchi, 2014). Despite high penetration and participation from RE investors in to FiT policy due to its luxury payment tariff offered, there are several challenges that need to be tackled to ensure the persistency of RE progress globally. The main challenges highlighted in general are (1) Tendency of having near-term upward pressure on electricity price that may leads to "windfall profits", (2) Wholesale electricity market price may be distorted, (3) Payment tariff for FiT are not rely from the current market price that makes the market for FiT not oriented, (4) FiT may exclude lower-income individuals from participating due to the high cost for installation and quota requirements, (5) Political interference and information discrepancy between public and private sector in FiT determination (Toby et al., 2010). Expertise and special monitoring on the FiT's policy implemented, actual cost and market integration are critical to ensure its compatible validity.

The first adoption of NEM was in the US state of Idaho (1980) and initiated with the Energy Policy Act 2005 (Rose and Meeusen, 2006). Since then, the United States has been recognized as a common model for NEM deployment that is applicable at the domestic level as well as in the industrial and commercial sectors. In certain countries such as Malaysia, NEM seemed to be a wise alternative to the FiT scheme. With current major transition from conventional grid into an intelligent grid, NEM's framework seems to be an appropriate way to stimulate the path of the future grid (Schelly, Louie, & Pearce, 2017). To be exact, NEM permits consumers to be prosumers (produce and consume electricity generated). The utility provider will execute as a coordinator providing electricity to consumers when production is less than demand or by enabling electricity injection into grid when there is a surplus of electricity generation. In return, consumers will receive payment for every excess of kW/H generated, thus resulting in savings in electricity bills since renewable energy credits will be accredited in the consumer's account toward the next billing cycle (Poullikkas, Kourtis, & Hadjipaschalis, 2013).

In addition to the many US states that have deployed NEM, a small number of countries in the EU—i.e., Belgium, Cyprus, Denmark, Italy, Greece, Portugal, and the Netherlands—are deploying it as well. NEM's scheme was observed to have an enormous successful at the high electricity rates and solar PV potential area. However, this scheme also possesses its own challenges especially on the variations of NEM policies deployed globally and overage tariff credits to consumers that below the current retail price (Koumparou et al., 2017). Home solar PV investors will only have positive net present value (NPV) if the electricity sold is above the levelized cost of electricity (Comello & Reichelstein, 2017). The challenge is to earn grid parity or positive NPV as a bonus. It is essential to ensure that investors attained adequate return on the project considering upfront capital expenditures, taxes applied and operating costs. The policy makers and government at the NEM adopted countries have to figure out on how to overcome these challenges since the FiT schemes in most countries globally (i.e., Germany) offered high FiT rates compared with its retail price in order to promote the deployment of RE.

Throughout the years, the NEM scheme has endured several revisions and improvements to its implementation in order to ensure parity benefits between utilities and consumers. Starting NEM at 2013, Las Vegas, for instance, has enrolled its NEM program with total incentives provided that amounted to USD 185 million after 1 year of deployment. Later, the Nevada Public Utilities Commission (PUC) engaged Energy, Environmental and Economics (E3) to conduct a cost−benefit analysis on NEM's current implementation (Price et al., 2014). Despite many benefits reported, the PUC observed that some variations are needed in the metering mechanisms to ensure the reaffirmation of NEM and at the same time to be less reliant on renewable portfolio standards (RPS). Together with NV Energy, they have initiated separate customers class for ratepayers, usage of time of use rates to preserve consumer NPV, and new consumer compensation schemes based on solar-specific service charges for fixed costs, demand charges for distribution costs, and energy charges (Davies & Carley, 2017). Comprehending both policy frameworks is an imperative in promoting RE deployment.

8.3.3 Monetary incentives

In terms of monetary incentives, substantial subsidies have been allocated by the government and private sectors in flourishing the development and production of renewable energy source electricity. Well-designed policies and the enforcement mechanism are essential elements to assist the financial incentives for RE project development while ensuring the project risks (IEA-RETD, 2017). The total global investment for fuel and RE amounted to USD 279.8 billion in 2017, with a 63% share coming from RE, and China alone marked as a lion's share of RE with 45%. Solar PV and wind power hold major contribution to RE development with 57% and 38%, respectively. Rapid developments of RE have directly increased the value of subsidy allocated, whereby at the latest, about USD 140 billion has been subsidized for RE power generation developments, though this value is still small compared with that of fossil fuel subsidies which are double those of RE (REN21, 2018).

Developed countries are continuing to revise their RE policies to suit the subsidy needed. In Japan, the Ministry of Economy, Trade and Industry is the government agency responsible for the enforcement of economic instruments policy. There are a few subsidies for RE still in force in Japan. One is the Subsidies of Local Projects for Accelerating the Introduction of New and Renewable Energy. This scheme was effective starting in 2000 for projects that contribute to the increment of growth of RE especially, wind power and solar PV. It followed the plans made by Japanese government in entitling this scheme, with total subsidy for this program amounting to 1232 million yen (IEA, 2000a). Another subsidy that is still in force is the Subsidy of Company for the Diffusion of Renewable Energy. Designated under the heating and cooling sector, this scheme was made effective in 2000 to speed up the diffusion of new and renewable energy in Japan. This scheme is eligible for companies involved in the utilization of new and renewable energy. For example, this subsidy will be given to companies that use gas-fired cogeneration with approximate funding estimated at 11.5 million yen (IEA, 2000b).

While it cannot be known for sure what subsidies are given for RE in Canada, one step in helping to move toward RE is that recently, the Canadian government promised to end fossil fuel subsidies no later than 2025 (Dolter & Rivers, 2018). Ending fossil fuel subsidies will force people to move toward RE. The best way for Canada to spur investment and keep Canada in the RE game is to set targets and goals for wind and solar power in the regional energy grid, says Merran Smith from Clean Energy Canada. In Smith's opinion, Clean energy doesn't need subsidies, it needs policies that commit to targets. A report by Chassin and Milke states that subsidies for green energy will be higher and cause demand for compensation for utilities as well as explicit subsidies.

8.3.4 Tax exemption

In conjunction with COP21 in 2015, the world united together to fight climate change by limiting the global warming to below 2°C. UNFCCC member states are committed to reducing world GHG emissions. Hence, members of UNFCCC that have signed the agreement are obliged to submit their transparency and accountable ambitious plans concerning GHG emission reductions every 5 years and to adhere to the proposal accordingly (UN, 2015c). Paving the path of climate mitigation plan, most countries globally are pursuing proactive actions such as the elimination of fossil fuel subsidies and ETS that reflect carbon pricing and subsidies granted for the purchasing and manufacturing of energy-efficient products and technologies (Carl & Fedor, 2016; Yuyin and Jinxi, 2018). Carbon pricing can be classified as carbon taxes and ETS. ETS is also recognized as a cap-and-trade system, baseline-and-credit system (Canada), and baseline-and-offset system (Australia) whereby carbon emissions are permitted to be traded. Carbon tax systems, defined as a fixed tax rate, implies that for the companies/manufacturers, carbon emissions and rates differ depending on the country (WBG, 2017). Canada, for instance, used CAD10/tCO_2 for any carbon production from fossil fuels in 2018, and the tax is expected to increase to CAD 50/tCO_2 by 2022. Through the literature, it can be seen that the initiation of carbon pricing has led to successful collection of more than USD $28.3 billion in revenue globally. The revenue is basically used to subsidize RE activities, energy efficiency, general funds, and direct rebates to individual or corporate taxpayers (Carl & Fedor, 2016).

After a century of dominance, fossil fuel power plants are no longer needed, whereby the total global energy generated by RES is anticipated to increase more than 60% by 2040 (IEA, 2017a). In promoting the same, several examples regarding taxes and incentives on RE will be discussed accordingly from the world's two largest economies, China and the United States. In China, the average CO_2 emitted annually is more than three billion tonnes. Therefore, Chinese policy has deployed carbon taxes and ETS to manage the taxes that will be applied due to the emission. The government of China, through its National Development and Reform Commission and National Energy Administration, had announced incentives of the usage of RES during peak demand and interprovince energy trade. Profits for electricity generated is now can be shared at a negotiated price between RE power producer and grid operators. Furthermore, to boost power carbon investment, China also implemented carbon pricing to include the cost of the environment following Western examples such as those of the United States and Denmark.

Several sources of RE financial funds are available in China, such as the National High Technology Research and Development Program of China ("863" Program), the National Basic Research Program of China ("973" Program), and the National Natural Science Fund. However, a major financial support provided by the Energy Bureau of National Development and Reform Commission is the Renewable Energy Development Special Fund (REDS fund), which is established to support various activities surrounding RE including resource exploration, standard-setting, and project demonstration. Since its announcement in 2015, REDS has reached 13.9 billion yuan of funds (Zhao, Chen, & Chang, 2016). Furthermore, the renewable energy law in China has stipulated to furnish favorable taxes for hydro, wind, solar, biomass, and geothermal power generation projects. Table 8.1 shows a list of available incentives in taxation available in China.

TABLE 8.1 China's fiscal and tax incentives for renewable energy.

Tax policies	RES element	Incentives significance	Ref.
Value added tax (VAT)	Solid waste (SW) energy recovery	Tariff exemption for SW that achieves the emission standard with no less than 80% of fuel power generation.	((MOF), 2008a)
	Methane biomass Small hydro Wind	13%, 6%, and 8.5% incentive given to the producer respectively from its 17% standard VAT.	(Zhao et al., 2016; (MOF), 2013)
	Solar	50% tax refund for solar power producer between from October 1, 2013, to December 31, 2015.	
Income tax	Hydro, Wind, Solar, Geothermal	Tax exemption on project development with capacity installed less than 250 MW for the first 3 years and reduction 50% on tax for 4–5 further years starting from initial project income from its 25% standard income tax.	((MOF), 2008b)
Import duty/ Import VAT	Wind	Effective January 1, 2008, refund tax used for R&D of new large-capacity wind power products is offered.	((MOF), 2008c)

Several incentives and tax exemptiond/rebated are available in the United States, such as the production tax credit (PTC), RPS, investment tax credit (ITC), and NEM. However, each that might be suitable for certain RES, such as ITC and NEM, are policies that have benefited solar power with perfomance-based incentives (PBIs) as the incentives offered. Approximately 20% of total electricity supply is generated from different sources of RE, which are hydropower, wind energy, biomass, and solar energy (Tran and Smith, 2017). Nevertheless, US electricity generation still possesses major share from fossil fuels. Motivated by the Paris Agreement, the United States is striving to realize its decarbonized target. However, in 2017 the federal government decided to withdraw from the agreement due to the domestic political matters and the president's own preferences (Zhang et al., 2017). Hence, it can be concluded that, apart from policy, technical, and economic matters, political interfererence may influence the energy transition.

Though the country has withdrawn from the Paris Agreement and has passive deployment of solar PV compared with that of European countries, it has nevertheless shown high penetration and development of solar PV over the last decade. ITC and NEM laws are used in the United States to accelerate the momentum of solar PV deployment (Stokes and Breetz, 2018). As for motivation, PBIs are introduced whereby actual energy production originated from solar PV is paid based on USD $/kWh. The common type of PBIs is FiTs. A solar system that is paid through a PBI approach must have metering installed that measures the output of the system, and this information is provided to the program administrator.

The California Solar Initiative is a PBI approach whereby incentives are paid monthly over a 5-year period (60 total payments) based on the actual energy (kWh) produced by a solar energy system. The incentive rate (USD $/kWh) should remain constant for the term of the contract. Applying PBI policy within the RE sector has motivated solar installers and system possessors to focus on proper installation, maintenance, and performance of their systems—since the payment is based upon the actual energy produced. This provides policy makers and regulators some assurance that incentives provided are being effectively managed and not squandered on systems with poor performance.

As for electric vehicles (EVs), 438 incentives, tax rebates, and sales tax subsidies are available. California's Zero-Emission Vehicle (ZEV) was introduced in 1990 toward the Clean Air Act's ozone standards with federal tax incentives offered to support EV deployment in general (Collantes & Sperling, 2008). The California Air Resources Board (CARB) has been appointed to assure ZEV achievement by demanding that car manufacturers build and sell EVs. After several years of revising the exact percentage of ZEV deployment, CARB announced that a 15% ZEV mandate is to be achieved starting from 2012 until 2025. Considering that, tax incentives such as the Alternative Motor Vehicle tax credit are applied to promote the usage of EV and hybrid vehicles (Milne, 2012).

Extensive financing for wind energy projects in the United is being boosted by the enacted policies of the production tax credit (PTC) and RPS. The adjustment through the Bipartisan Budget Act of 2018 claimed that in 2017, PTC has underlined the inflation value in USD $ 0.019/kWh for wind electricity generated facilities. However, the value varies for subsequent years by about 20%, 40%, and 60% PTC off for wind facilities commencement in 2017, 2018, and 2019 respectively (DSIRE, 2018). The policies under RPS are responsible for ensuring that the electricity supplier contributes some percentage of electricity production from RE regardless of the type of source—solar, geothermal, wind, or biomass. Despite many arguments arising about exact percentages and participants, approximately 68% of wind energy projects in 37 states were able to meet the policy targets (Barbose, 2016).

8.3.5 Impacts of renewable energy policy

Several policies have been developed and enforced in order to encourage the adoption of RE throughout the world. Basically, almost all countries will have similar policies enacted, but specific policies are essential to suit respective national contexts and overwhelm possible market barriers and failures (Felix Groba, 2013). Accordingly, the policies introduced should cover incentives, tax credits, emissions trading systems (ETS), RE quota requirements, and pricing laws. FiT and RPS are the most popular and useable mechanisms under RE policy. Given the appropriate design features, Denmark and Germany possess greater experience with FiT, while for the United Kingdom it is with RPS. The effective policies will determine the pace of RE development as well as provide positive impacts in terms of industrial development, job creation, and investment (Lipp, 2007).

Rapid development and deployment of RE has had various remarkable impacts worldwide, such as global competitiveness in RE industries, falling cost of RE tenders; market growth of the electrification and digitization of transport; reduced fossil fuel subsidies; minimizing environmental impact; and continuously new policies, initiatives, and missions set by groups of governments at all levels. The world is united to battle climate change through scaling up deployment of efficient, modern, and sustainable low-carbon technology options. In order to keep the momentum of RE development, sources of funding must be sufficient. Revenue obtained by the fossil fuel tax, carbon pricing, and ETS will be used to support RE technologies and carbon reduction goals. To date, China has world's largest ETS, with its first phase of cap-and-trade program concentrated on the power sector (UNFCCC, 2017).

However, despite positive feedback on RE policies, several unanticipated impacts have occurred in several countries. The abundance of PV installed in China has caused individual investors and RE farms to experience delayed payments due to the limitations on cash flow and uncertain payment timelines by the national grid and policies. Conflict occurred in Indonesia in the form of rivalry between institutions in procuring RE premises and the adjacent location of hydropower plants that were built, due to a frenzy to acquire the most profitable sites. Those undesired impacts must be scrutinized, rectified, and included in the policy design to encourage continued strong growth in RE development that could benefit environmental, economic, and energy security. Nevertheless, tangible policies and financial and political supports are crucial to ensure the ambitious targets for green pathways that are achievable.

References

Africa, E. (April 11 , 2018). *Africa's energy revolution is underway, but challenges remain.* https://www.esi-africa.com/africas-energy-revolution-is-underway-but-challenges-remain/(27/08/2018).

Australian Department of Industry and Science. (2015). *Energy White Paper.* Retrieved from https://apo.org.au/node/54017.

Avila, N., et al. (2017). *The energy challenge in sub-saharan Africa: A guide for advocates and policy makers: Part 1: Generating energy for sustainable and equitable development.* Oxfam Research Backgrounder series. https://www.oxfamamerica.org/static/media/files/oxfam-RAEL-energySSA-pt1.pdf.

Barbose, G. (2016). *U. S. Renewables Portfolio standards 2016 annual status report.* Lawrence Berkeley National Laboratory.

Barrington-Leigh, C., & Ouliaris, M. (2017). The renewable energy landscape in Canada: A spatial analysis. *Renewable and Sustainable Energy Reviews, 75*, 809–819.

Buxton, N. (2016). COP 21 charades: Spin, lies and real hope in Paris. *Globalizations, 13*(6), 934–937.

Carl, J., & Fedor, D. (2016). Tracking global carbon revenues: A survey of carbon taxes versus cap-and-trade in the real world. *Energy Policy, 96*, 50–77.

Chen, W.-M., Kim, H., & Yamaguchi, H. (2014). Renewable energy in eastern Asia: Renewable energy policy review and comparative SWOT analysis for promoting renewable energy in Japan, South Korea, and Taiwan. *Energy Policy, 74*, 319–329.

Collantes, G., & Sperling, D. (2008). The origin of California's zero emission vehicle mandate. *Transportation Research Part A: Policy and Practice, 42*(10), 1302–1313.

Comello, S., & Reichelstein, S. (2017). Cost competitiveness of residential solar PV: The impact of net metering restrictions. *Renewable and Sustainable Energy Reviews, 75*, 46–57.

Corporation, K. E. M. (2011). *South Korean new and renewable Energies (NRE).*

Davies, L. L., & Carley, S. (2017). Emerging shadows in national solar policy? Nevada's net metering transition in context. *The Electricity Journal, 30*(1), 33–42.

DIICCSRTE. (2013). *Department of Industry, Innovation, Climate Change, Science, Research and Tertiary Education (2013). How Australia's carbon price is working, one year on: Department of Industry, Innovation. Climate Change, Science, Research and Tertiary Education.* https://trove.nla.gov.au/work/186836713?q&versionId=203384913.

Dolter, B., & Rivers, N. (2018). The cost of decarbonizing the Canadian electricity system. *Energy Policy, 113*, 135–148.

DSIRE. (2018). *Database of State Incentives for Renewables & Efficiency, Renewable Electricity Production Tax Credit (PTC).* https://programs.dsireusa.org/system/program/detail/734.

EC. (2009). *European Commission, 2020 Climate & Energy Package - Climate Changes.* Retrieved from https://ec.europa.eu/clima/policies/strategies/2020_en.

EC. (2015). *European Commission, European development policy, 2015*. Available from https://ec.europa.eu/europeaid/policies/european-development-policy/2030-agenda-sustainable-development_en.

EC. (2017). *European Commission, Energy union and climate, 2017*. Available from https://ec.europa.eu/commission/priorities/energy-union-and-climate_en.

Eurostat. (2018). *Share of energy from renewable sources in gross final consumption of energy 2004–2016*. Available from http://ec.europa.eu/eurostat/statistics-explained/images/b/b7/Table_2-Share_of_energy_from_renewable_sources_in_gross_final_consumption_of_energy_2004-2016.png.

Faure-Schuyer, A., Welsch, M., & Pye, S. (2017). Chapter 7 – A market-based European energy policy. In *Europe's energy transition - insights for policy making* (pp. 41–48). Academic Press.

Felix Groba, B. B. (2013). *Impact of renewable energy policy and use on innovation: A literature review*. DIW Berlin, German Institute for Economic Research.

Foundation, A.-E. (2017). *Asia-europe energy policy forum*. Available from http://www.asef.org/index.php/projects/themes/sustainable-development/4204-asef-side-event-at-cop23-.

Goldthau, A. (2016). *The handbook of global energy policy* (1 edition). Wiley-Blackwell. November 14, 2016.

Hasanuzzaman, M., Rahim, N. A., Hosenuzzaman, M., Saidur, R., Mahbubul, I. M., & Rashid, M. M. (2012). Energy savings in the combustion based process heating in industrial sector. *Renewable and Sustainable Energy Reviews, 16*(7), 4527–4536.

IEA RETD TCP. (2017). *Fostering Renewable Energy integration in the industry (RE-INDUSTRY), IEA RE Technology Deployment Technology Collaboration Programme (IEA RETD TCP), Utrecht, 2017*. http://iea-retd.org/wp-content/uploads/2017/03/RE-INDUSTRY-Final-report-1.pdf.

IEA. (2000a). *International Energy Agency. Subsidies of local projects for accelerating the introduction of new and renewable energy, 2000, Subsidies of Local Projects for Accelerating the Introduction of New and Renewable Energy*. Retrieved from https://www.iea.org/policiesandmeasures/pams/japan/name-21252-en.php.

IEA. (2000b). *International Energy Agency. Subsidy of Companies for the Diffusion of Renewable Energy*. Retrieved from https://www.iea.org/policiesandmeasures/pams/japan/name-21251-en.php.

IEA. (2011). *International Energy Agency. New and Renewable Energies (NRE)*. Retrieved from https://www.iea.org/policiesandmeasures/pams/korea/name-25045-en.php.

IEA. (2014). *International Energy Agency. National Energy Policy (Government Regulation No. 79/2014)*. Retrieved from https://www.iea.org/policiesandmeasures/pams/indonesia/name-140164-en.php.

IEA. (2017a). *International Energy Agency. Energy Supply and Consumption Revolution Strategy (2016-2030)*. Retrieved from https://www.iea.org/policiesandmeasures/pams/china/name-162879-en.php.

IEA. (2017b). *International Energy Agency. World Energy Outlook 2017 (WEO-2017)*. Retrieved from https://www.oecd.org/about/publishing/Corrigendum_EnergyOutlook2017.pdf.

IEA. (2018). *International Energy Agency. Energy Policies Of IEA Countries - Australia*. Retrieved from https://www.connaissancedesenergies.org/sites/default/files/pdf-actualites/australia2018.pdf.

Ilham, N., et al. (2018). A review on comparison of technologies and progress of a smart grid development in Malaysia and Singapore. *Journal of Fundamental and Applied Sciences, 10*(6S), 1323–1337.

IqtiyaniIlham, N., Hasanuzzaman, M., & Hosenuzzaman, M. (2017). European smart grid prospects, policies, and challenges. *Renewable and Sustainable Energy Reviews, 67*, 776–790.

Kester, J., III, Moyer, R., & Song, G. (2015). Down the line: Assessing the trajectory of energy policy research development: Research in energy and natural resource policy. *Policy Studies Journal, 43*(S1), S40–S55.

Kinley, R. (2017). Climate change after Paris: From turning point to transformation. *Climate Policy, 17*(1), 9–15.

Koumparou, I., et al. (2017). Configuring residential PV net-metering policies – a focus on the Mediterranean region. *Renewable Energy, 113*, 795–812.

Kumar Singh, B. (2013). South Asia energy security: Challenges and opportunities. *Energy Policy, 63*, 458–468.

Lipp, J. (2007). Lessons for effective renewable electricity policy from Denmark, Germany and the United Kingdom. *Energy Policy, 35*(11), 5481–5495.

Mendonça, M., & Corre, J. (2009). *Success story: Feed-in tariffs support renewable energy in Germany.* Available from http://www.e-parl.net/eparliament/pdf/080603%20FIT%20toolkit.pdf.

Milne, J. E. (2012). Electric vehicles: Plugging into the US tax code. In *Green taxation and environmental sustainability.* Edward Elgar.: Edward Elgar.

MoF. (2008a). *Ministry of Finance, China. Notice on the comprehensive utilization of resources and value-added tax policy for other products, July 2008.* Retrieved from http://www.mof.gov.cn/index.htm.

MoF. (2008b). *Ministry of Finance, China. The preferential enterprise income tax catalogue for public infrastructure projects (the 2008 edition), September 2008.* Retrieved from http://www.mof.gov.cn/index.htm.

MoF. (2008c). *Ministry of Finance, China. Notice on import tax policies of high-power wind turbine and its key components and raw materials, April 2008.* Retrieved from http://www.mof.gov.cn/index.htm.

MoF. (2013). *Ministry of Finance, China, . Notice on VAT of Solar PV Power Generation Products (September 2013).* Retrieved from http://www.mof.gov.cn/index.htm.

Momoko, A., et al. (2017). *Economic and energy outlook of Japan through FY2018.* Japan: Japan: The Institute of Energy Economic.

Morgan, J. (2016). Paris COP 21: Power that speaks the truth? *Globalizations, 13*(6), 943–951.

Niesten, E., Jolink, A., & Chappin, M. (2018). Investments in the Dutch onshore wind energy industry: A review of investor profiles and the impact of renewable energy subsidies. *Renewable and Sustainable Energy Reviews, 81*, 2519–2525.

Park, J., & Hong, T. (2013). Analysis of South Korea's economic growth, carbon dioxide emission, and energy consumption using the Markov switching model. *Renewable and Sustainable Energy Reviews, 18*, 543–551.

Parliament of Australia. (2010). *Emission Trading Scheme.* Retrieved from https://www.aph.gov.au/About_Parliament/Parliamentary_Departments/Parliamentary_Library/Browse_by_Topic/ClimateChangeold/responses/economic/emissions/emissionstrade (12/07/2015).

Pearse, R. (2016). The coal question that emissions trading has not answered. *Energy Policy, 99*, 319–328.

Poruschi, L., Ambrey, C. L., & Smart, J. C. R. (2018). Revisiting feed-in tariffs in Australia: A review. *Renewable and Sustainable Energy Reviews, 82*, 260–270.

Poullikkas, A., Kourtis, G., & Hadjipaschalis, I. (2013). A review of net metering mechanism for electricity renewable energy sources. *International Journal of Energy and Environment, 4*(6).

Price, S., et al. (2014). *Nevada net energy metering impacts evaluation.*

Pyrgou, A., Kylili, A., & Fokaides, P. A. (2016). The future of the Feed-in Tariff (FiT) scheme in Europe: The case of photovoltaics. *Energy Policy, 95*, 94–102.

REN21. (2018). *Renewable 2018 - Global Status Report*. Retrieved from http://www.ren21.net/wp-content/uploads/2018/06/17-8652_GSR2018_FullReport_web_-1.pdf.

Ritchie, H., & Roser, M. (2018). *Fossil Fuels*. https://ourworldindata.org/fossil-fuels.

Rose, K., & Meeusen, K. (2006). *Reference manual and procuderes for implementation of the "PURPA standards" in the Energy Policy Act 2005*. Available from https://www.energy.gov/sites/prod/files/Manual%20for%20Implementation%20of%20PURPA%20Standards%20in%20EPACT%202005%20%28March%202006%29.pdf.

Sahoo, S. K., et al. (2016). Energy efficiency in India: Achievements, challenges and legality. *Energy Policy, 88*, 495–503.

Schelly, C., Louie, E. P., & Pearce, J. M. (2017). Examining interconnection and net metering policy for distributed generation in the United States. *Renewable Energy Focus, 22–23*, 10–19.

SEDA. (2016). *SUSTAINABLE ENERGY DEVELOPMENT AUTHORITY (SEDA) MALAYSIA. Malaysia Feed in Tariff and Net Energy Metering*. Retrieved from http://www.seda.gov.my.

Shahbaz, M., et al. (2013). Economic growth, energy consumption, financial development, international trade and CO_2 emissions in Indonesia. *Renewable and Sustainable Energy Reviews, 25*, 109–121.

Sovacool, B. K. (2014). Energy studies need social science. *Nature, 511*, 529–530.

Stokes, L. C., & Breetz, H. L. (2018). Politics in the U.S. energy transition: Case studies of solar, wind, biofuels and electric vehicles policy. *Energy Policy, 113*, 76–86.

Swartz, J. (2016). China's national emissions trading system: Implications for carbon markets and trade. In *ICTSD global platform on climate change, trade and sustainable energy; climate change architecture series*. Geneva, Switzerland: International Centre for Trade and Sustainable Development.

Tang, C. F., Tan, B. W., & Ozturk, I. (2016). Energy consumption and economic growth in Vietnam. *Renewable and Sustainable Energy Reviews, 54*, 1506–1514.

The 13th five-year plan of China. (2016). *The 13th five-year plan: for economic and social development of the people's republic of china (2016-2020)*. Central Compilation & Translation Press. http://en.ndrc.gov.cn/newsrelease/201612/P020161207645765233498.pdf (22/07/2019).

TIA. (2011). *Telecommunications Industry Association, Smart Grid Policy Roadmap: Consumer Focused and Technology Driven*. Retrieved from https://www.tiaonline.org/wp-content/uploads/2018/05/Smart_Grid_Policy_Roadmap_-_Consumer_Focused_and_Technology_Driven.pdf.

Toby, D.,C., et al. (2010). *A policymaker's guide to feed-in tariff policy design*.

Tran, T. T. D., & Smith, A. D. (2017). fEvaluation of renewable energy technologies and their potential for technical integration and cost-effective use within the U.S. energy sector. *Renewable and Sustainable Energy Reviews, 80*, 1372–1388.

UN. (2015). *United Nations, Paris agreement*. Available from https://treaties.un.org/pages/ViewDetails.aspx?src=TREATY&mtdsg_no=XXVII-7-d&chapter=27&clang=_en.

UNFCCC. (2017). *China to launch world's largest emissions trading system, 2017*. Available from https://unfccc.int/news/china-to-launch-world-s-largest-emissions-trading-system.

United Nations. (2017). *The Sustainable Development Goals Report*. Retrieved from New York https://unstats.un.org/sdgs/files/report/2017/TheSustainableDevelopmentGoalsReport2017.pdf.

WBG (2017) World Bank, Ecofys and Vivid Economics. (2017). *State and Trends of Carbon Pricing 2017 (November), by World Bank, Washington, DC*. https://doi.org/10.1596/978-1-4648-1218-7.

Webber, M. E. (2016). *Energy 101: Energy, technology, and policy*. Texas University of Texas Press.

Yuyin, Y., & Jinxi, L. (2018). The effect of governmental policies of carbon taxes and energy-saving subsidies on enterprise decisions in a two-echelon supply chain. *Journal of Cleaner Production, 181*, 675–691.

Zhang, H.-B., et al. (2017). U.S. withdrawal from the Paris agreement: Reasons, impacts, and China's response. *Advances in Climate Change Research, 8*(4), 220–225.

Zhao, Z.-Y., Chen, Y.-L., & Chang, R.-D. (2016). How to stimulate renewable energy power generation effectively? – China's incentive approaches and lessons. *Renewable Energy, 92*, 147–156.

Index

Note: 'Page numbers followed by "f" indicate figures and "t" indicate tables'.

F

G

H

I

R

S

T

U

V

W

Z

Printed in the United States
By Bookmasters